Stereochemical and Stereophysical Behaviour of Macrocycles

stereochemistry of organometallic and inorganic compounds 2

Stereochemical and Stereophysical Behaviour of Macrocycles

Edited by

IVAN BERNAL

Department of Chemistry, University of Houston, Houston TX 77004, U.S.A.

Elsevier

Amsterdam — Oxford — New York — Tokyo 1987

ELSEVIER SCIENCE PUBLISHERS B.V.
Sara Burgerhartstraat 25
P.O. Box 211, 1000 AE Amsterdam, The Netherlands

ELSEVIER SCIENCE PUBLISHING COMPANY INC.
52 Vanderbilt Avenue
New York, NY 10017, U.S.A.

ISBN 0-444-42815-1 (Vol. 2)
ISBN 0-444-42604-3 (Series)

© Elsevier Science Publishers B.V., 1987

Printed in The Netherlands

Introduction to Volume 2

Volume 1 appeared in the Spring of 1986, was twice reviewed in journals and a number of times in oral or written private communications. There is general agreement that the series will serve a valuable role in surveying complex subjects for readers who need the unifying synthesis only an expert can provide; however, as expected, there is no overall consensus as to what should be the content of each individual volume. Most commentators seem to favor dedication of each tome to a single topic-- a few spoke favorably on behalf of variety. Many understand the problem of filling an entire volume with equally relevant and timely material and all understand the problem of gathering, at a given point in time, a group of experts who have the time and inclination to write a survey of their specialties, even when, admittedly, those specialties are ripe for evaluation.

Nonetheless, an effort was made to provide a sharper focus in the content of Volume 2 which will deal exclusively with coordination compounds of macrocyclic ligands. The three chapters deal with the relationship between stereochemistry and other chemical and physical properties of metal macrocyclics. The rationale for this selection is stated in Chapter 1 by Boeyens and Dobson: "the interest in metallic macrocyclics, as reflected in the current literature, establishes this field of study as one of the most prolific topics in coordination chemistry." Their review was largely and "arbitrarily restricted to the stereochemistry of transition-metal macromonocycles, for ring sizes not exceeding sixteen, and excluding crown ethers."

The relationship between thermodynamics and stereochemistry of macrocyclics and cryptates is explored in Chapter 2 by Buschmann, who remarks that "only one year after Werner discovered chelate compounds [1893, ed.] the first crown ether was synthesized by Vorländer"; yet the rate at which synthetic work has progressed in recent years is such that "in 1985 Izatt and co-workers tabulated the known thermodynamic data for cation-macrocyclic interactions (ref. 19). At that time the complexation properties of almost 260 different ligands were given. Thus, the properties of 90% of the known ligands have not yet been estimated."

Matthes and Parker review, principally, stereochemical aspects of the macrocycles of second and third row transition elements with additional comments on the stereochemistry of Cu and Ni in unusual oxidation states and their survey, though somewhat briefer in appearance, covers, as do all others, information published well into 1986 when the first drafts of manuscripts were due.

We hope this volume elicits interest in the topics covered among those not familiar with the subject reviewed and serves as an update for those active in the field.   Our aim is to provide an educational vehicle for the neophyte and an update for those actively participating in the growth of the discipline, and your comments will   inform us of the degree to which these expectations are met.

Many thanks to those who have taken the time to privately convey, orally or in writing,   their reaction to Volume 1 and for suggestions as to topics we should cover,   and by whom. Such information is invaluable to us and, to the extent that we can persuade suitable reviewers to contribute, efforts will be made to cover those topics if, in the estimation of the potential reviewer, they have not recently been treated elsewhere. Volumes 3 to 5 are already committed and, to no small degree, will reflect the judicious choice of people who cared to make suggestions of topical material which interests them.   Heartfelt thanks to them, also.

A letter by Professor John C. Bailar, Jr., commenting on a chapter in Volume 1,   was particularly satisfying and I share it with the chemical community so as to preserve a bit of history of American chemistry. Among other comments, he remarked that (and I quote with his kind permission):

"Jackson has done an excellent job on the inversion reactions [i.e., The Bailar Inversion Reaction; ed] and he is to be commended for it.   He obviously has a tremendously wide knowledge of all the reactions involved. When Auten and I observed the reaction in 1934 I did not realize it would precipitate such a great amount of subsequent research, and that it would lead to new ideas about reaction mechanisms.   Incidentally this was the first piece of work which I did in coordination compounds, and my organic chemical background, of course, played an important role in it.   It seemed quite reasonable to me that since   Walden has observed products of different chirality by using potassium hydroxide and silver hydroxide that we would find comparable results in the inorganic field.   Auten, who was an undergraduate student at that time, had serious doubts about it and I think was a little loath to try the experiment.   Naturally we were delighted when we found that there is an inversion in the reaction which we studied."

TABLE OF CONTENTS

CHAPTER 1

STEREOCHEMISTRY OF METALLIC MACROCYCLICS

J.C.A. Boeyens and S.M. Dobson

CHAPTER 2
THERMODYNAMIC AND STEREOCHEMICAL ASPECTS OF THE
MACROCYCLIC AND CRYPTATE EFFECTS
H.-J. Buschmann

CHAPTER 3
STEREOCHEMICAL ASPECTS OF MACROCYCLIC COMPLEXES OF
TRANSITION METAL IONS
K.E. Matthes and D. Parker

CONTRIBUTORS TO THIS VOLUME

J.C.A. Boeyens and S.M. Dobson
Structural Chemistry Group
Department of Chemistry
University of the Witwatersrand
1 Jan Smuts Avenue
Johannesburg
South Africa

H.-J. Buschmann
Physikalische Chemie
Universität-GH Siegen
Postfach 101240
D-5900 Siegen
Federal Republic of Germany

K.E. Matthes and D. Parker
Department of Chemistry
University of Durham
South Road
Durham DH1 3LE
United Kingdom

CHAPTER 1

STEREOCHEMISTRY OF METALLIC MACROCYCLICS

JAN C.A. BOEYENS and SUSAN M. DOBSON

Structural Chemistry Group, Department of Chemistry,
University of the Witwatersrand,
1 Jan Smuts Avenue, Johannesburg, South Africa.

STEREOCHEMISTRY OF METALLIC MACROCYCLICS

1. INTRODUCTION

The interest in metallic macrocyclics, as reflected in the current literature, establishes this field of study as one of the most prolific topics in coordination chemistry. It deals with the encirclement of a metal ion by a cyclic ligand, which seems to require a minimum ring size of about nine atoms. All compounds of interest are organic heterocycles containing at least three heteroatoms like N, P, O or S. The crown ethers belong to this class, but distinguish themselves by their marked ability to complex alkali metal ions. All the ligand atoms in crown complexes are oxygen by definition, and because of the physiological importance of the alkali ions these compounds are of special biological significance. For this reason, crown ethers tend to be more comprehensively and regularly reviewed than the nitrogen, phosphorus, sulphur and mixed-donor macrocyclics. The emphasis is reversed in this review.

The special relative disposition of cation and ligand in macrocyclics generates properties distinct from those of regular coordination compounds. The macrocyclic effect refers to their special stability and is but one, even though the best documented example. No discussion of the exceptional properties of macrocyclics can therefore be complete without taking proper account of stereochemical factors.

1.1 HISTORICAL SURVEY

The earliest known examples of metallic macrocyclics were found in naturally occurring biologically active substances like hemoglobin, chlorophyll and vitamin B_{12}, containing Fe, Mg and Co respectively. The unadorned macrocyclic moieties, basic to these substances have skeletons of the type shown in Figure 1. A synthetic isomeric variety thereof has recently been described (1).

In this type of structure a planar configuration is dictated by the extensive delocalization and supported by the fused five-membered rings. Without these features, the molecular conformation of macromonocycles is a more flexible function of ring strain, degree of unsaturation and the nature of the central metal ion. Stereochemical variety will be limited only by the ingenuity of the synthetic chemist.

Figure 1. Basic skeleton of naturally occurring macrocyclics.

The first synthetic metallic macrocycle was reported by Curtis (2) and in the intervening two-and-a-half decades, a large variety of macrocycles have been synthesized and their complexation chemistry has been studied extensively. By analogy with the known natural products most of the early synthetic materials were tetradentate nitrogen donor complexes. For a macrocycle of this type to encircle a first-row transition metal ion, it must consist of 13 to 16 members and the spacing between donor atoms must be such that 5, 6 or 7 membered chelate rings are formed (3). For a smaller ring or a larger metal ion the macrocycle has to fold, in order to attain maximum coordination.

An obvious method to produce macromonocycles is by cyclization of linear polydentate ligands. Comparison of the metal complexes of the open-chain and cyclic analogues show that the cyclization has a dramatic effect on complex properties. This is typically manifested as

(i)   an increased kinetic inertness towards complex formation in aqueous medium and towards decomposition (4);

(ii) increased ligand-field strength (5);

(iii) an increase by orders of magnitude in thermodynamic stability (6);

(iv) improved stabilization of high metal ion oxidation states (7).

These factors collectively define the well-known macrocyclic effect (6).

Various aspects of macrocyclic chemistry have been reviewed in the literature. Gokel and co-workers (8) reviewed the synthesis of aliphatic nitrogen-donor macrocyclics. Christensen and co-workers (9) reviewed the synthesis of multidentate macrocyclics and their ion-binding properties. Izatt and co-workers (10) collated and reviewed the kinetic and thermodynamic data on cation-macrocycle interactions, and Lindoy (11) discussed the transition metal macrocyclics.

1.2 CHEMICAL CLASSIFICATION

The chemical identity of a macromonocyclic ligand depends on the ring size and the number, nature and location of the donor atoms and double bonds. The IUPAC rules (12) for naming organic compounds result in the assignment of unequivocal, but extremely complicated names to these compounds and a more popular procedure is to introduce some abbreviated version. Most widely accepted is the set of rules proposed by Melson (13). In its simplest form it consists of a number specifying the ring size, a term denoting degree of unsaturation and the symbols of ligating atoms in alphabetical order. The numbering of ring atoms starts with the heteroatom of highest priority according to Chemical Abstracts rules, i.e. $O > S > Se > N > P > As > Sb$, etc., and proceeds in the direction that yields the lowest locants for all heteroatoms and sites of unsaturation. The rules are illustrated by the following examples:

$[9]$ane-4,7-N_2-1-S

or

$[9]$ane-N_2S

11,13,13-Me_3-$[13]$-10ene-1,4,7,10-N_4

or

$Me_3[13]$ene-N_4

$5,7,7,12,14,14,-Me_6-[14]-4,11-diene-1,4,8,11-N_4$

or

$Me_6[14]4,11-dieneN_4$

1.3   MOLECULAR CONFORMATION

The idea of molecular shape is an integral part of chemical thinking (14). It features at all explanatory levels, shaping important concepts like chemical reactivity, reaction mechanism, phase transition and reaction pathway. This is often collectively referred to as the structure-function relationship, although molecular structure is rarely invoked explicitly. It enters the argument in the guise of concepts like steric congestion, ligand-field stabilization, barriers to conversion, overcrowding, packing considerations, cone-angles, Jahn-Teller distortion, trans effects, steric control, vibronic coupling, transition states, reaction intermediates, and many other terms, making brief appearances in the literature, only to explain some poorly characterized observations. This diversity of concepts have one feature in common. They cannot be reduced to fundamental electronic factors and represent some manifestation of molecular conformation. There is nothing amiss with this, provided the fundamental importance of molecular shape is appreciated. This requires a primary ability to identify conformational classes and provide a logical basis for structure-function relationships.

The purpose of this review is to outline the stereochemical classification of a typical family of macrocyclic metal complexes, which has once before been classified in terms of composition and some ad hoc conformational features (15). The present attempt approaches the classification on the basis of more firmly defined principles.

A first look at molecules differentiates between configurations that are topologically one, two or three dimensional, representing classes of string-like, ring-like and thing-like structures. The compounds of interest in this review are of the two-dimensional type, with a prominent, large heterocycle, that encircles a metal ion, as their most characteristic feature. It is noticed at the outset that the presence of the metal ion must exert a decisive influence on the configuration of the macrocycle and conformational types not represented among the free ligands can be expected to occur. A classification of the conformational types adopted by the organic moiety in macrocyclic complexes should therefore look beyond the standard low-energy configurations assumed by the free ligands. Existing classifications as developed for large-ring systems would therefore not be suitable without modification, for the description of macrocyclic complexes. Although final interest would focus on large-ring systems the conformational principles are best introduced by a consideration of small rings.

2. THE CONFORMATION OF CHEMICAL RINGS

The conformation of any cyclic molecule is fully described by the set of endocyclic torsion angles. According to the definition of Klyne and Prelog (16) the torsion angle about the bond jk in the fragment ijkl is the angle between the projections of the bonds ij and kl on to a plane perpendicular to the bond jk when viewed in the direction from j to k. Its value is positive if a clockwise rotation brings the projection of bond ij to coincide with the projection of kl. It is noted that the sign does not change with the point of view. This sign convention is illustrated by the following simple examples, showing the signs of torsion angles about the central bonds:

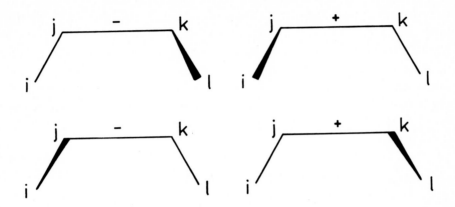

The sign relationship between symmetry-related bonds is readily established on the basis of these examples. The effect of a two-fold axis, a mirror plane (m ≡ $\bar{2}$) and a centre of symmetry ($\bar{1}$) is illustrated below.

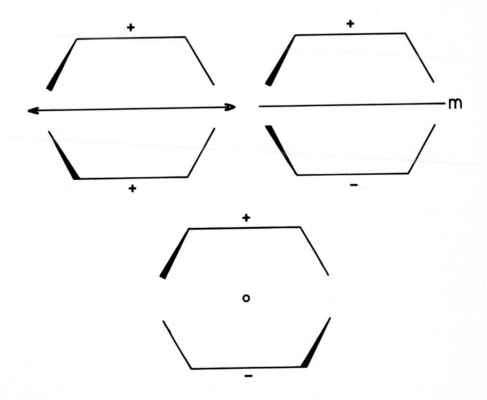

Bonds related by rotation axes have identical torsion angles and those related by inversion axes have equal torsion angles with opposite signs.

The orientation of any bond with respect to those adjacent to it is usually obvious from the sign of its torsion angle. Considered in clockwise sense a positive sign indicates an increasing elevation, as demonstrated for a puckered four-membered ring, below.

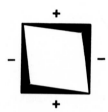

The torsion angles for a ring system of known symmetry can therefore readily be specified in terms of the foregoing simple principles. Conversely, the symmetry or conformational type of a ring system can be determined qualitatively from a set of experimentally measurable endocyclic torsion angles. However, the actual conformations of cyclic molecules only rarely correspond to exact symmetrical types and a more quantitative approach to identify intermediate types is required.

The number of possible conformational types of cyclic compounds increases with ring size and flexibility. The one conformational type common to all cyclics is the planar arrangement, which is recognized by a set of zero endocyclic torsion angles. Any puckered ring has some non-zero torsion angles and the degree of pucker is related to the magnitude of these. Ring conformation can therefore be defined in terms of the number and identity of non-zero torsion angles and their associated amplitudes of puckering. A systematic analysis of all possible combinations of non-zero (positive and negative) torsion angles therefore defines the complete field of allowed conformations for a ring of any size. Algorithms to analyse the conformations of small ring systems, with less than nine members, exist and in principle any experimentally encountered shape can be reduced to a linear combination of a small number of symmetrical types. This procedure is illustrated in the next paragraph.

2.1   SMALL RING SYSTEMS

All schemes for the specification of nonplanarity of closed rings in cyclic compounds derive from the basic formula of Kilpatrick and others (17) to describe the out-of-plane displacement of the jth atom, perpendicular to the plane of a five-membered ring:

$$z_j = (2/5)^{\frac{1}{2}} q \cos[2\psi + 4\pi (j-1)/5] \tag{1}$$

where q is a puckering amplitude and ψ is a phase angle which describes various kinds of puckering. Motion involving a change in ψ at constant q is described as pseudorotation. This approach was generalized in terms of cartesian atomic coordinates for rings of any size and varying bond lengths by Cremer and Pople (18). The mean plane is defined by setting the sum of the perpendicular atomic displacements,

$$\sum_{j=1}^{N} z_j = 0$$

and fixed by the conditions of zero angular momentum:

$$\sum_{j=1}^{N} z_j \begin{Bmatrix} \cos \\ \sin \end{Bmatrix} \left[2\pi(j-1)/N \right] = 0$$

To obtain expressions of type (1) a set of generalized puckering parameters is defined by expressions of the type

$$q_m \begin{Bmatrix} \cos \\ \sin \end{Bmatrix} \phi_m = \pm(2/N)^{\frac{1}{2}} \sum_{j=1}^{N} z_j \begin{Bmatrix} \cos \\ \sin \end{Bmatrix} \left[2\pi m(j-1)/N \right]$$

for $N > 3$ and $2 \leq m \leq (N-1)/2$.

For odd-membered rings there are $(N-3)/2$ amplitude-phase pairs (q_m, ϕ_m) and even-membered rings have $(N-4)/2$ pairs and an additional amplitude, $q_{N/2}$. A total puckering amplitude is obtained as

$$Q = \left(\sum_m q_m^2 \right)^{\frac{1}{2}} = \left(\sum_j z_j^2 \right)^{\frac{1}{2}}$$

The smallest ring with pucker is N=4 with equal atomic deviations from the mean plane, $z_1 = -z_2 = z_3 = -z_4$, and the puckering amplitude $Q = \pm 2z_1$.

For a five-membered ring there is a single amplitude-phase pair (q, ϕ) and the displacement expression becomes

$$z_j = (2/5)^{\frac{1}{2}} q \cos[\phi + 4\pi(j-1)/5].$$

For a twist ring with a two-fold axis, through atom 1, $q \cos \phi = (2/5)^{\frac{1}{2}} z_1 = 0$, since the dyad lies in the mean plane, and it follows that $\phi = 90°$ or $270°$. If the two-fold axis is moved to coincide with atoms 2, 3, etc., the phase angle changes by steps of $\pi/5 = 36°$. Symmetrical envelope conformations can be shown to occur at $\phi = n\pi/5$, $n = 0 \rightarrow 10$. It follows that ten twist and ten envelope conformations occur at alternate values of $n\pi/10$ from $\phi = 0$ to $360°$ on a pseudorotational cycle. The conformational symbol at each stage depends on the atomic numbering. Using superscripts and subscripts for unique atoms, above and below the mean plane respectively, the scheme defines the pseudorotational cycle shown in Figure 2.1. It is noted that anti-clockwise numbering effectively inverts the ring through the mean plane, advances the phase angle by 2π and generates the enantiomeric conformational symbol. For a planar ring the amplitude $Q = 0$. A widely used convention in coordination chemistry describes the helicity of a five-membered chelate ring as either λ or δ if the phase angle $\phi < 180°$ or $\phi > 180°$, respectively, and the atomic numbering starts at the metal ion.

For non-planar six-membered rings it is convenient to convert the parameters of pucker into a set of polar coordinates (19) Q, θ and ϕ by introducing an angle θ to define the ratio between the amplitudes q_2 and q_3.

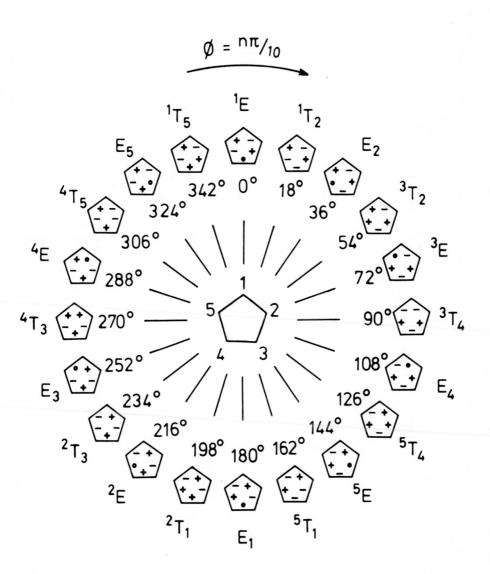

Figure 2.1. The pseudorotational cycle of five-membered rings

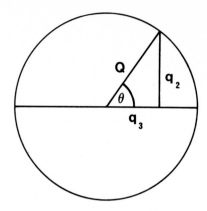

These coordinates can be used to map the conformation of a ring on the surface of a sphere. Since, by definition

$$q_3 = N^{-\frac{1}{2}} \sum_{j=1}^{6} (-1)^{j-1} z_j$$

it assumes a value of zero for symmetrical arrangements that produce the same number of positive and negative displacements from the mean plane. This implies $\cos \theta = 0$ and hence these arrangements occur along the equator ($\theta = 90°$) on the surface of a sphere to define a pseudorotational cycle with intervals $\phi = n\pi/6$. As in the rotational cycle for five-membered rings, conformations with two-fold and mirror symmetry, respectively, alternate around the circle. These are six twist and six boat conformations. A two-dimensional polar projection of the spherical surface, with all the symmetrical conformations of six-membered rings drawn in (20) is shown in Figure 2.2. Starting from a chair conformation, which occurs at the poles, an inversion pathway is seen to occur via intermediates which retain either two-fold or mirror symmetry. The two-fold pathway goes via half-chair (H), screw-boat (S) and twist (T) conformations and the mirror pathway goes via envelope (E) and boat (B) forms. The two polar angles can be used to read off the conformation of a general ring with non-zero Q directly from Figure 2.2.

Puckered seven-membered rings are characterized by the two amplitude-phase pairs q_2, ϕ_2 and q_3, ϕ_3, which can geometrically be associated with the surface of a variable torus (21). The detailed mapping of all conformational types on a toroidal surface

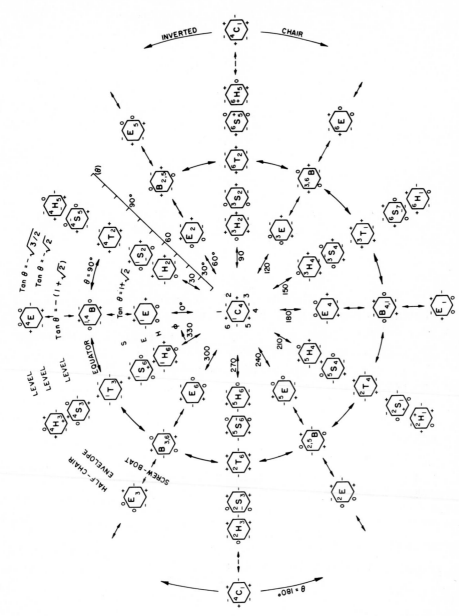

Figure 2.2. Two-dimensional polar projection of a spherical surface to distinguish between the symmetrical conformations of six-membered rings

has been established (22) and is summarized in Figure 2.3, which represents a polar projection of a torus along the major radial track.

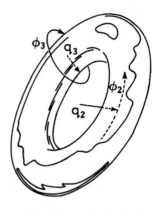

The Xn and Sn forms represent configurations with a mirror plane through atom n and the opposite bond, whereas Kn and Tn represent configurations with a two-fold axis in the same way. Different forms of the same symmetry class can clearly project at the same point which is defined only by the angular coordinates. By use of the symbols B, C, S and T for boat, chair, sofa and twist configurations respectively the various symmetry sets consist of:

X = C, B, BS
T = TB, TS
K = TC
S = S

The conformations of these various types are defined below, where in all cases n = 1 and with clockwise atomic numbering starting at the top.

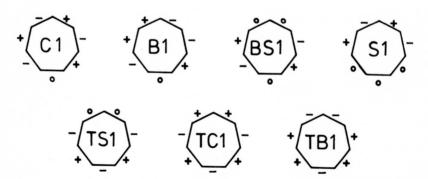

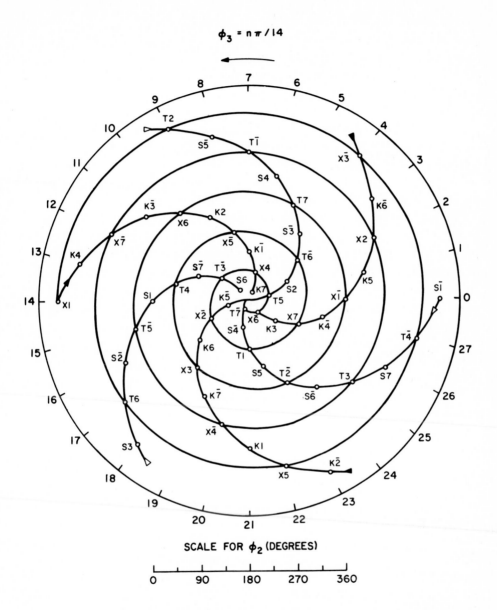

Figure 2.3. Polar projection of a torus along the major radial track. Pseudorotational and deformation pathways appear as spirals. The symbols (X), (T) and (K) represent the forms (C, B, BS), (TB, TS) and (TC) respectively. The phase angle ϕ_2 is measured radially from the spiral origin on a linear scale

A negative sign (bar) attached to a symbol in Figure 2.3 indicates
the inverse form of a conformation, i.e., with the signs of all
endocyclic torsion angles inverted. To identify the actual
conformation at an ambiguous site from Figure 2.3, it is necessary
to take into account the values of the puckering coordinates,
which can be specified as an angle $\theta = \tan^{-1}(q_2/q_3)$, as before.

The boat-twist boat pseudorotational forms have $\theta = 90°$,
whereas the chair-twist chair forms occur at $\tan \theta = 1/\sqrt{2}$. The
three sofa forms, BS, S and TS have $\tan \theta = 5/\sqrt{2}$, 1 and $1/\sqrt{2}$
respectively. Overlapping forms are therefore clearly distin-
guished by θ.

An eight-membered ring is considered to be the largest of the
small rings. In principle its conformation can be characterized
in terms of three puckering amplitudes and two phases, as was done
for smaller rings. Although most of the interconversions have
been established by Hendricksen (23) a detailed protocol to
identify the various forms on the basis of puckering parameters
has not been established yet. Since eight-membered rings are of
little importance in the present context they will not be
discussed any further.

The approach to conformational analysis of small rings,
outlined here, can formally be extended to larger ring systems.
However, the number of forms and of puckering parameters becomes
excessive and of little interpretational value. A different
approach is required for the large macrocycles, but the system of
smaller rings, containing the metal ion, can be usefully analysed
in terms of the concepts already developed.

2.2   LARGE RINGS

The most promising scheme for the stereochemical characteriza-
tion of large-ring systems was proposed and developed by Dale
(24). An earlier attempt to rationalize the conformations of
macrocylics was described by Bosnich et al. (25). They described
the possible conformations for tetraazamacrocyclics of various
sizes on the basis of amine hydrogen atom, or substituent,
positions with respect to the macrocyclic plane. This approach is
too restrictive for the present purpose. It can even be extended
to oxygen and sulfur atoms with reference to their remaining lone-
pairs, but that would still confine the discussion to tetradentate
macrocyclics.

The Dale approach is based on the most likely disposition of neighboring bonds. The accepted terminology to distinguish between various arrangements of bonds in hydrocarbon chains is illustrated by the Newman projection along the bond. Positions 1, 2 and 3 with respect to R, define torsion angles of 60, -60 and 180° respectively. These low-energy arrangements are called gauche (+), gauche (-) and anti respectively. The fully eclipsed, or syn arrangement is the energetically least favourable with a somewhat lower barrier at the ϕ = 120°, eclipsed arrangement.

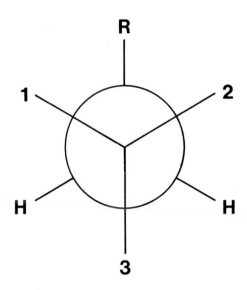

In large rings there is sufficient freedom for the energetically best conformation to include only relatively staggered bonds, with the largest possible number in anti form. For rings to exist, a number of gauche bonds must inevitably occur. The problem is to establish the smallest number of gauche bonds with the best distribution in terms of relative position and signs of the torsion angles. The resulting conformation is then described in terms of the number and position of bends, or corners introduced to form a ring. The first requirement is therefore to identify bond sequences that produce bends. Some of the possibilities considered are shown diagramatically below:

18

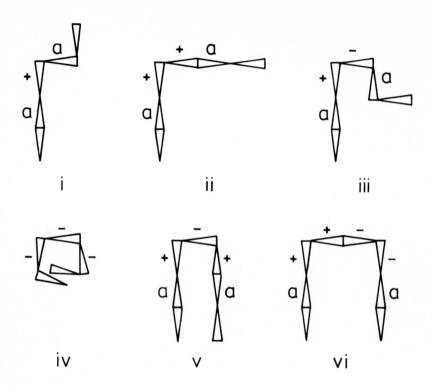

i   ii   iii

iv   v   vi

(i)   An isolated <u>gauche</u> bond is considered not efficient for a bend, becoming of interest only for rings larger than 16-membered.

(ii) A sequence of two isolated <u>gauche</u> bonds with the same sign is efficient as a bend.

(iii) Two consecutive <u>gauche</u> bonds of opposite sign is considered to be excluded by large steric interactions in open chains and hence should be of little importance in ring formation.

(iv) A sequence of three or more gauche bonds leads to a helical structure.

(v)   A sequence of three <u>gauche</u> bonds of alternating sign has inherent strain, that could be relieved by systematic adjustment of the torsion angles and should therefore occur as an efficient double bend in ring formation.

(vi) A sequence of four <u>gauche</u> bonds is of interest only when they occur as two pairs with opposite signs, defining two bends of type (ii).

This scheme recognizes a two-fold axis through a bond, in an odd-membered ring as defining a corner, equivalent to a normal ++ or -- bend. Corresponding conformations with mirror planes are excluded on steric grounds. To specify the resulting conformations a shorthand notation is introduced. It consists of a series of numbers within brackets, enumerating the number of bonds between bends, starting from the smallest. The direction around the ring is chosen to also minimize the next number in the sequence.

To apply the rules it is necessary to relax the definition of the gauche arrangement so as to include torsion angles that are different from 60°, due to systematic adjustment. To achieve consistency Dale proposed that all torsion angles, $|\phi| \leq 120^\circ$ should be included, and with this stipulation managed to assign unambiguous symbols to most low-energy conformations of cycloalkanes, from cyclononane to cyclohexadecane.

This scheme can clearly not be extended to metallic macrocyclics, without modification. When a macrocycle binds to a metal, additional rings are formed, usually five- and six-membered rings. These chelate rings impose additional steric demands and the macrocycle could adopt a conformation that would be energetically unfavourable in the cycloalkane analogue. Further constraints arise from the condition of sufficient overlap between donor-atom lone pairs and suitable metal orbitals.

2.3 MACROCYCLIC RINGS

When applied to macrocyclics it is necessary to modify and extend the Dale (24) scheme in order to ensure a minimum of three corners for any of these cyclic arrangements. The term "non-angular conformation", introduced by Dale does not seem to serve any useful purpose. It is in fact only necessary to add an isolated gauche bond and a gauche pair with opposite sign, as possible bends, to establish a classification, valid for all macrocyclics. The wider definition of a gauche arrangement ($|\phi| < 120^\circ$) to allow for steric modifications is probably too liberal, in that it produces too many one-bond sides and hence unwieldy conformational formulae. The rules for corner definition therefore reduce to:

(i)   A corner occurs at the junction of any two <u>gauche</u> bonds, irrespective of sign, e.g.:

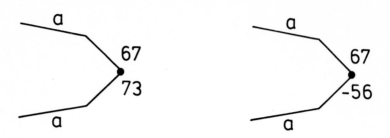

where any torsion angle, a of $|\phi| > 90°$ is considered <u>non-gauche</u>. This 90° discrimination between anti and gauche forms is consistent with the IUPAC rules for describing the stereochemistry of conformational isomers, Rule E(23) (26). A symmetrical arrangement of this kind has a two-fold axis or mirror plane, respectively, at the corner, Figure 2.4

(ii)   A corner exists at the junction of an isolated <u>gauche</u> bond and the adjacent bond with the smaller $|\phi|$.

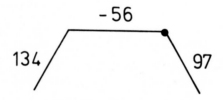

(iii) When an isolated gauche bond contains a two-fold or pseudo two-fold axis it has a corner on either side.

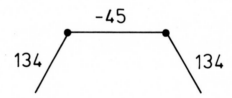

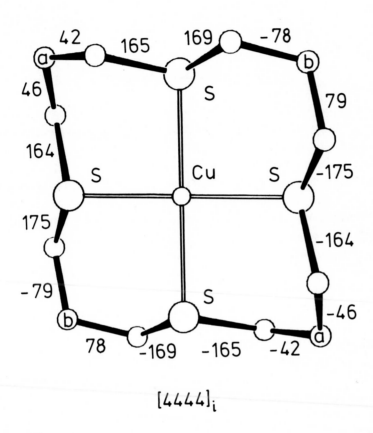

$$[4444]_i$$

Figure 2.4. The endocyclic torsion angles in the [16]ane-S_4 copper(II) complex showing the Dale type corners (a) as well as the additional corners defined in this work (b)

A mirror or pseudo mirror plane of this type, e.g.

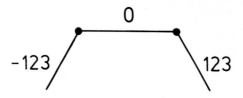

is excluded on steric grounds.

  (iv) Two adjacent one-bond sides coalesce into a single two-bond side.

  Stereochemical descriptions of all macrocyclics in the remainder of this work are based on this classification scheme.

3. THE STRUCTURE OF MACROCYCLICS

  Macrocyclic chemistry has developed largely within the era of fast computerized crystallographic analyses and the structures of all new compounds have been determined almost routinely by diffraction methods.

  There is one serious complication with the crystallographic analysis of macrocyclics, arising from molecular disorder. This is related to the disc-like structure, even of puckered macrocyclic rings. One finds a minimum in packing energy for more than one orientation of a ring, within its mean plane. Crystallographic superposition of the two or more identical molecules, by projection into a single unit cell, from a random distribution, then introduces extra symmetry. The additional elements of symmetry could be either of a local nature, or appear as crystallographic symmetry. The latter case is the more lethal as it can easily be overlooked and the symmetry of disorder mistaken for molecular symmetry, with obvious disastrous consequences. Examples of the different types of disorder are briefly discussed below.

  An additional symmetry element introduced by rotational disorder of a macrocyclic ring can be a mirror plane, a two-fold axis, or a centre of symmetry. A remarkable consequence of two-fold disorder is the observed isomorphism of the two isomeric perchlorate complexes of nickel [14]ane-1,4,7,11-N_4 and [14]ane-1,4,8,11-N_4 in the space group Pna2$_1$ (27). Resolution of the disorder was achieved by rigid-body refinement of trial structures, obtained by molecular mechanics simulation (28). The disordered individuals were found to be related by local two-fold axes as illustrated in Figure 3.1.

The picture is even more confusing in the structure of the nickel perchlorate complex of [9]ane-N_2O (29). In this case a mirror plane, generated by the disorder is also an element of crystallographic and apparent molecular symmetry. A chemically impossible average structure of 2/m symmetry, Figure 3.2, is obtained by crystallographic analysis. Once again, the disorder can be resolved by molecular mechanics and shown to arise from the superposition of rotationally distinct individuals.

A chemically suspect mirror plane also occurs in the chromium isocyanate complex of [15]ane-1,4,8,12-N_4 (30) in the space group Pnam. This disorder problem was partially overcome by refinement, with constraints in space group $Pna2_1$. It actually appears that many other 15-membered macrocyclics are also disordered, with the metal ion at a centre of symmetry (31, 32, 33) or on a two-fold axis (34). It is of interest to note that this mode of disorder does not occur in macrocyclics with thirteen-membered rings. This probably relates to the planar coordination in the fifteen-membered macrocyclics, compared to a non-planar arrangement in the thirteen-membered analogues.

The more specific structural features of macrocyclics are conveniently separated into aspects related to ring conformation and coordination geometry respectively. The latter aspect is deferred to a later chapter and the former will be discussed in relation to ring size. The ease of metal-ion encapsulation by monocyclic macrocycles is a critical function of ring size. The smaller rings cannot encircle a metal ion, and form macrocyclic complexes of the sandwich type. An intermediate ring size can accommodate a metal ion close to its mean plane, at a distance depending on ionic size. For large-ring systems the metal ion lies comfortably in the plane of the ring. The structures of metallic macrocyclics are therefore conveniently discussed within these categories.

3.1 SMALL-RING MACROCYCLICS

The smallest compounds of this class are the nine-membered macrocyclics. Details of published crystallographic analyses of

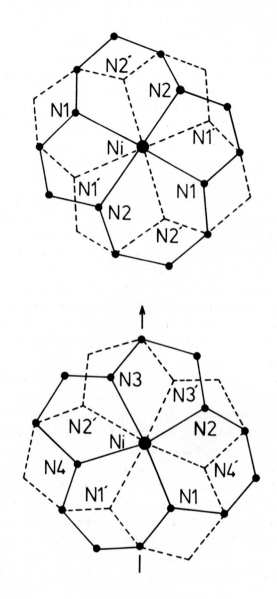

Figure 3.1. The disordered structures in isomorphous crystals of
[Ni(cyclam)]$^{2+}$ (27) and [Ni(isocyclam)]$^{2+}$ (28)

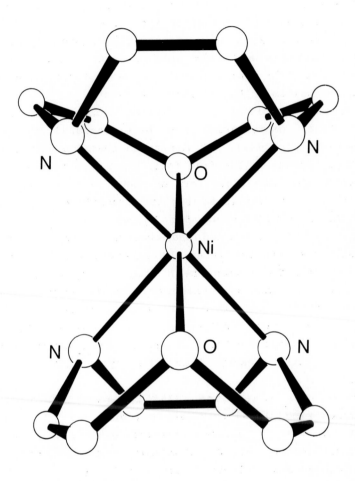

Figure 3.2. Chemically impossible average structure of symmetry 2/m derived for the $[Ni([9]ane-N_3)_2]^{2+}$ cation by non-critical crystallographic analysis

these compounds are summarized in Table 1. Only saturated ring compounds with three donor atoms in the ring are represented. The nine-membered ring adopts either the [333] or the [234] conformations. The helicity of the five-membered chelate rings in the [333] class is either ($\lambda\lambda\lambda$) or ($\delta\delta\delta$) and ($\lambda\delta\delta$) or ($\delta\lambda\lambda$) for [234]. Since, in all cases the metal ion is above the macrocyclic plane and coordinated by the donor atom lone pairs, conformations like [522] and [12222] cannot occur. The highest symmetry that occurs for [333] is a 3-fold axis through the cation. It is noteworthy that this applies, with few exceptions, whenever the coordinated metal is in the +3 oxidation state.

The relative stability of different cyclononanes in either [333] or [234] conformation has been compared by molecular mechanics simulation (42). It was found that for all S-N mixed-donor [9]ane ligands in the free form, the [234] configuration was more strained than the [333] form by about 12 - 20 kJ mol^{-1}. The same difference is calculated for the transition-metal complexes of mixed-donor ligands, but for N_3 and S_3 macrocyclics the difference is more pronounced and amounts to about 35 kJ mol^{-1}. The calculations also show that the expected increase in complex stability is not found when one N atom is replaced by S. This implies that the loss of symmetry in going from [9]ane-N_3 to [9]ane-N_2S partially cancels out the favourable decrease in steric energy due to the replacement of nitrogen by sulfur. Unsymmetrical ligands can therefore adopt the lower-symmetry [234] conformation more cheaply than the symmetrical N_3 and S_3 macrocycles.

The difference between the [333] and [234] conformations of nine-membered rings is illustrated in Figure 3.3. The [234] form is allowed for ligands that can interact asymmetrically with metal ions which cannot accept symmetrical facial coordination. Examples include Jahn-Teller distorted Cu(II) (41) and Pt(II) that forms a square-planar complex with [9]ane-N_3 (46), functioning as a bidentate ligand, as shown in Figure 3.3.

Most of the complexes have distorted octahedral coordination geometries since the tridentate ligand is generally facially coordinating. This holds for mono, di, tri, tetra and octanuclear complexes. The exceptions are the Pt(II) complex, already mentioned, the Cu(II) complex discussed below, the Mo(II) complex where because of the larger cationic size seven coordination

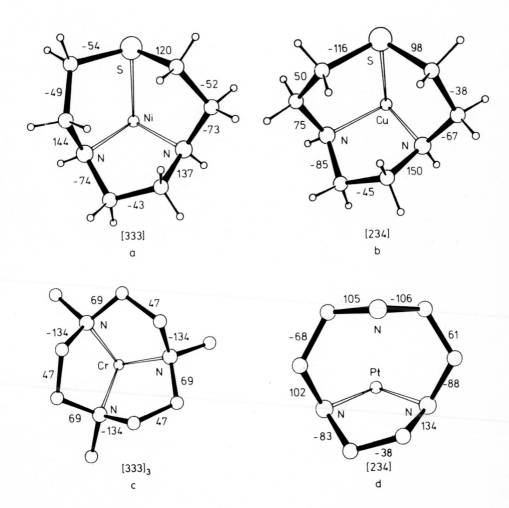

Figure 3.3. Macrocycles with [333] and [234] conformations in the complexes: a) [Ni([9]ane-N_2S)$_2$](NO$_3$)$_2$ (40), b) [Cu([9]ane-N_2S)$_2$](NO$_3$)$_2$ (41), c) [Cr(Me$_3$-[9]ane-N_3)]$_2$(-OH)$_3$ (68) and d) [Pt([9]ane-N_3)$_2$]Cl$_2$, (46)

TABLE 1
Crystallographically characterized metal complexes of 9-membered macrocyclic ligands

Macrocycle	Other ligands	Cation	Cat:Macr. Ratio	Coordination		Macrocyclic Conformation	Ref.	Remarks
				No.	Geometry			
[9]ane-S_3		None				$[333]^a$	35	
		Ni(II)^h	1:2	6	Octahedral^b	[333]	36	M-S bond lengths increase in the order Co<Ni<Cu
		Cu(II)	1:2	6	Tetragonal	[333]	36	
		Co(II)^l	1:2	6	Octahedral	[333]	36	
		Fe(II)^l	1:2	6	Octahedral	[333]	37	
		Co(III)	1:2	6	Octahedral	[333]	38	
	(CO)$_3$	Mo(0)	1:1	6	Octahedral	[333]	39	
[9]ane-N_2S		Ni(II)^h	1:2	6	Octahedral	[333]	40	
		Cu(II)	1:2	6	Tetragonal	[234]	41	NO_3^- H-bonded to macrocycle
	Br$_2$	Cu(II)	1:1	5	Square pyramid	[234]	42	Unpublished
[9]ane-N_3		Ni(II)^h	1:2	6	Octahedral	[333]	43	Trigonal distortion
	Br$_2$	Cu(II)	1:1	5	Square pyramid	[333]	44	

[9]ane-N$_3$	Cl$_2$	Cu(II)	1:1	5	Square pyramid		45	
		Pt(II)	1:2	4	Square planar	[234]	46	One uncoordinated Nitrogen
NH$_3$, C$_2$H$_4$NO$_2$		Co(II)	1:2	6	Octahedral	[333]	38	
		Co(III)	1:1	6	Octahedral	[333]	47	
		Fe(III)	1:2	6	Octahedral	[333]$_3$	48	
		Fe(II)	1:2	6	Octahedral	[333]	48	
		Ni(III)	1:2	6	Tetragonal	[333]	49	
(CO)$_3$, Br		Mo(II)	1:1	7	Piano Stool	[333]	50	
H$_2$O, trans (μ-OH)$_2$		Co(III)	2:2	6	Octahedral	[333]	51	
H$_2$O, trans (μ-OH)$_2$		Rh(III)	2:2	6	Octahedral	[333]	52	
μ-CO$_3^{2-}$, (μ-OH)$_2$		Cr(III)	2:2	6	Octahedral	[333]	53	
μ-O, (μ-CH$_3$CO$_2^-$)$_2$		Fe(III)	2:2	6	Octahedral	[333]	54	
(μ-OH)$_2$, trans O		Mo(V)	2:2	7	Irregular	[333]	55	
(μ-O)$_2$, cis O		Mo(V)	2:2	7	Irregular	[333]	55	
(μ-OH)$_2$, trans Cl		Mo(III)	2:2	6	Octahedral	[333]	56	
(μ-OH)$_2$, μ-CH$_3$CO$_2^-$		Mo(III)	2:2	6	Octahedral	[333]	56	
(μ-OH)$_2$, trans O		V(IV)	2:2	6	Octahedral	[333]	57	Planar Mo–O–Mo / O fragment

(continued)

Ligand	Bridge	Metal	Ratio	CN	Geometry	Symmetry	Ref.	Notes
[9]ane-N$_3$	(μ-OH)$_2$, μ-CH$_3$CO$_2^-$	Ru(III)	2:2	7	Irregular	[333]	58	
	μ-Br, μ-OH, transBr	Mo(III)	2:2	7	Irregular	[333]	59	
	μ-O, (μ-CH$_3$CO$_2^-$)$_2$	Mn(III)	2:2	6	Octahedral	[333]	60	
	(μ-OH)$_2$, μ-OH, OH	Cr(III)	3:3	6	Octahedral	[333]$_3$	61	
	(μ-O)$_3$	Mn(IV)	4:4	6	Octahedral	[333]	62	
	(μ$_3$-O), (μ$_2$-OH)	Fe(III)	8:6	6	Octahedral	[333]	63	
	(O)$_3$	Re(VII)	1:1	6	Distorted Octahedral	[333]$_3$	64	
	(μ-OH)$_3$	In(III)	4:4	6	Distorted Octahedral	[333]	65	
	(CH$_3$CO$_2^-$)$_2$ (μ-O)	In(III)	2:2	6	Distorted Octahedral	[333]	65	
	(ClO$_4^-$)$_3$	Pb(II)	1:1	6	Distorted Octahedral	[333]	66	
	(NO$_3$)$_3$	Pb(II)	1:1	6	Distorted Octahedral	[333]	66	
[9]ane-N$_2$O		Ni(II)[h]	1:2	6	Octahedral	[333]^d	29	
2-Me-[9]ane-N$_3$		Co(III)	1:2	6	Octahedral	3-fold^d	67	Disorder unresolved
N-Me$_3$-[9]ane-N$_3$	μ-O, (μ-CH$_3$CO$_2^-$)$_2$	Mn(III)	2:2	6	Octahedral	D$^d_{2h}$	60	
	(μ-OH)$_3$	Cr(III)	2:2	6	Octahedral	[333]$_3$	68	
	(μ-OH)$_3$	Rh(III)	2:2	6	Octahedral	[333]$_3$	68	

N-Me$_3$-[9]ane-N$_3$ μ-O,O2	Mo(VI)	2:2	6	Octahedral	[333]	69	
(μ-HNCN)$_2$	Cu(II)	2:2	5	Trigonal bipyramid	d	70	
(μ-N$_3$)$_3$	Ni(II)	2:2	6	Octahedral	[333]	71	
	Tl(I)	1:1	6	Distorted Octahedral	[333]	72	
TACNTA	Cr(III)	1:1	6	Octahedral	[333]	73	
	Fe(III)[h]	1:1	6	Trigonal	[333]$_3$	73	Three pendant groups
	Cu(II)	1:1	6	Trigonal prismatic	[333]	73	
	Ni(III)	1:1	6	Octahedral	[333]$_3$	74	
N-(ES)$_3$ [9]ane-N$_3$	Ni(II)	1:1	6	Octahedral	[333]	75	
H$_2$O	Cu(II)	1:1	5	Square pyramid	[333]	75	
N-(AE)$_3$ [9]ane-N$_3$	Co(III)	1:1	6	Octahedral	[333]$_3$	76	

a — molecular symmetry is given as a subscript to the conformation type
b — octahedral does not imply full octahedral symmetry, likewise for other geometries
d — disorder
h — high spin
l — low spin
ES — 2-ethanesulfonate
AE — 2-aminoethyl
TACNTA — 1,4,7-Triazacyclononane-N',N'',N'''-triacetate

becomes possible and those complexes of Mo and Ru where the coordination geometry is disturbed by metal-metal bonding.

1,4,7-Triazacyclononane-N,N',N''-triacetate (TACNTA), which is [9]ane-N_3 with 3 pendant acetate groups, can force the metal ion towards trigonal prismatic geometry in order to function as a hexadentate ligand. In the case of the Fe(III) and Cu(II) complexes (73) the coordination sphere is closer to a trigonal prism than an octahedron.

In the Cu(II) macrocyclics of the type [Cu(mac)X_2] where mac = [9]ane-N_3 or [9]ane-N_2S and X = Br or Cl square pyramidal five coordination is observed. This also occurs in the trisethanesulfonate N-substituted [9]ane-N_3 derivative, Na[Cu(mac)].3H2O (75). The actual coordination geometry of the dimeric complex [Cu_2(Me_3-[9]ane-N_3)$_2$(NCN)$_2$]$^{2+}$ is related to a trigonal bipyramid, but this could be an artifact of disorder. In only two structures is 6-coordinate Cu(II) sandwiched between two macrocycles, i.e., [Cu([9]ane-SX_2)]$^{2+}$ where X = S or N. In the S_3 complexes there is surprisingly little tetragonal distortion (0,04Å) of the coordination octahedron. This still awaits satisfactory explanation and could relate to disorder.

As indicated already, disorder is an important factor in the analysis of macrocyclic conformation. In the present case, and for other odd-membered rings, a crystallographic mirror plane arising from disorder is easily recognized where it bisects a dimethyl bridge in a chemically unlikely eclipsed configuration (29). In other cases the space group requires higher symmetry than allowed by the known chemical structure of the molecule, like a three-fold axis for Me-[9]ane-N_3 (67). In all these cases anomalously high thermal vibration parameters and excessive molecular distortion usually provide another indication of disorder.

The structure of the only ten-membered macrocyclic that has been crystallographically characterized is shown in Figure 3.4. The macrocycle has the [2323] conformation that corresponds with the lowest energy conformation of cyclodecane as calculated by Dale (24). The six-membered chelate ring is in chair conformation (Puckering parameter, $\phi = 172°$) and the two five-membered rings both have ($\phi = 106°$ and $251°$ respectively) an envelope conformation.

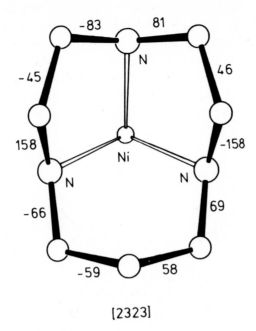

[2323]

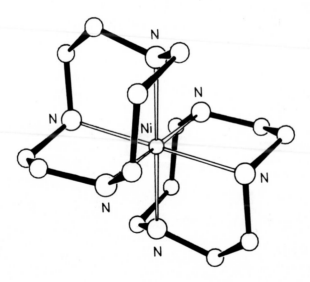

Figure 3.4. Macrocyclic conformation and coordination geometry of
    $[Ni([10]ane-N_3)_2]^{2+}$ (77)

3.2   MEDIUM-SIZED MACROCYCLICS

Twelve and thirteen-membered macrocyclic rings are too small to encapsulate most metal ions, but provided the lone pairs on the donor atoms are suitably oriented, stable 1:1 complexes are formed by tetradentate macrocyclic ligands. In the case of tetra-azacycloalkanes the orientation of the four NH groups with respect to the best plane through the macrocyclic ring defines the following basic situations:

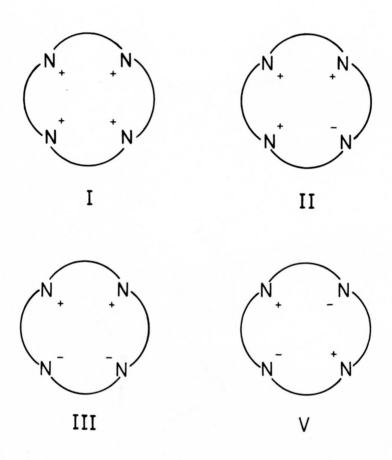

The + and - signs indicate the positions of H atoms relative to the flattened macrocyclic ring (25). Variations on the theme occur as the number of methylene groups between different N atoms changes. In the case of twelve-membered macrocycles each loop consists of two methylene groups to produce four five-membered chelate rings on complexation. In a thirteen-membered macrocyclic the siting of the six-membered chelate can produce different arrangements of classes II and III.

Twelve-membered macrocyclic rings can have two, three or four-fold axial symmetry. Examples of three-fold symmetry in tridentate phosphorus and nitrogen donor ligands are shown in Figure 3.5. The conformation in both cases is [444] and the main difference is in the size of the torsion angles around the donor atoms. The presence of this three-fold axis is easily confirmed by non-zero values of the q_3 and q_6 puckering amplitudes only. A twelve-membered ring with a perpendicular two-fold axis has non-zero puckering amplitudes q_2, q_4 and q_6. Figure 3.5 illustrates a four-fold axis in a nitrogen-donor twelve-membered macrocycle. The conformation is [3333] and only q_4 has a non-zero value.

Structural details of all crystallographically characterized twelve-membered macrocyclics are summarized in Table 2. Most of the macrocycles have four donor atoms and since the ligands cannot encircle metal ions, the most natural arrangement is five coordination from a type I macrocycle and a monodentate axial ligand. The coordination is either square pyramidal or trigonal bipyramidal (Figure 3.6) and the macrocycle has no preferred conformation. Tridentate ligands coordinate facially with conformation [2334] or [13233] (Figure 3.7). It is noted that all twelve-macrocycles occur with [3333] conformation as free ligands.

In the complexation of cations such as Co(III) and high-spin Ni(II) that prefer six coordination, the macrocycle must fold so as to leave two cis positions vacant for other ligands. This can be achieved by macrocycles of either type I or II structure, but the conformation is invariably [2424] (Figure 3.8). There seems to be a preference of type II structure in Co(III) complexes and type I for Ni(II). Although type I structures are compatible with the conformations [3333], [2343], [2424] and [2334], type II always requires the folded [2424] conformation.

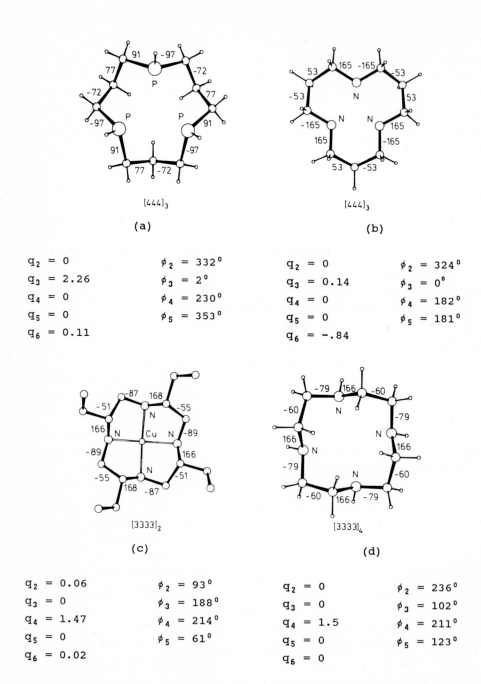

Figure 3.5. Endocyclic torsion angles and puckering parameters for 12-membered macrocycles with 3-fold (a and b), 2-fold (c) and 4-fold (d) axes

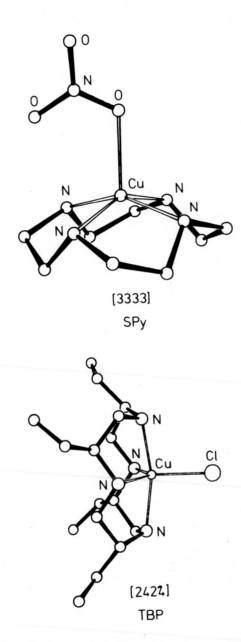

[3333]

SPy

[2424]

TBP

Figure 3.6. Cyclododecane complexes with Square Pyramidal and Trigonal Bipyramidal Coordination

TABLE 2
Crystallographically characterized metal complexes of 12-membered macrocyclic ligands

Macrocycle	Other ligands	Cation	Cat:Macr. Ratio	Coordination No.	Geometry	Macrocyclic Conf.	Type	Ref.	Remarks
[12]ane-N_3	NO_3^-,bidentate	Cu(II)	1:1	5	Trigonal bipyramid	[2334]		78	
	$(\mu-N_3)_2$,N_3	Ni(II)	2:2	6	Octahedral	[13233]		71	
[12]ane-P_3	$(CO)_3$	Mo	1:1	6	Octahedral	[13233]		79	
[12]ane-S_4	H_2O	Cu(II)	1:1	5	Sq.Pyramid	[2334]	I	32	
[12]ane-O_4		Ag(I)	1:2	8	Anti Prism	[3333]	I	80	
[12]ane-N_4	NO_3^-	Cu(II)	1:1	5	Sq.Pyramid	[3333]	I	81	
	$(NO_2)_2$	Co(III)	1:1	6	Cis-Octahedral	[2424]	II	82	
	β-diketone	Co(III)	1:1	6	Cis-Octahedral	[2424]	II	83	
	CO_3^{2-}	Co(III)	1:1	6	Cis-Octahedral	[2424]	II	84	
1,7-Me_2-[12]ane-N_4	Br^-,H_2O	Ni(II)[h]	1:1	6	Cis-Octahedral	[2424]	I	85	
	CO_3^{2-},bidentate	Co(III)	1:1	6	Cis-Octahedral	[2424]	II	86	

Ligand	Anion	Metal	Ratio	C.N.	Geometry	Pattern	Type	No.	Remarks
1,4,7,10-Me$_4$-[12]ane-N$_4$		Ni(II)	1:1	4	Sq.Planar	[2334]	I	87	Ni atom above coordination plane
2,5,8,11-Et$_4$-[12]ane-N$_4$	Br$^-$,OBr$_2^{2-}$	Co(III)	1:1	6	Cis-Octahedral	[2424]	II	88	N-chirality: SSSR C-Chirality: RRRR
	Cl$^-$	Cu(III)	1:1	5	Sq.Pyramid	[3333] [3333]	I I	89 90	Free ligand
1,4,7,10-(EtOH)$_4$-[12]ane-N$_4$		Li(I), Na(I), K(I)		5, 7, 8		[3333]	I	91 91 91	Hydrate with intramolecular O-H--N bonds
10,10'-ethylene bis[12]ane-NO$_3$						[3333]	I	92	Free ligand
N-(acetato)$_4$ [12]ane-N$_4$	Acetates H$_2$O	Cu(II)	2:1	6	Cis-Octahedral	[2424]	I	93	Second Cu(II) in binuclear complex with normal carboxylate structure
	H$_2$O	Eu(III)	1:1	9	Capped anti Prism	[3333]	I	94	

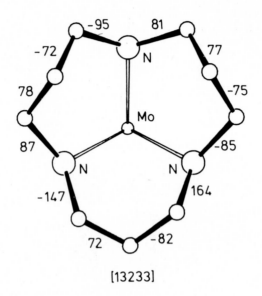

[13233]

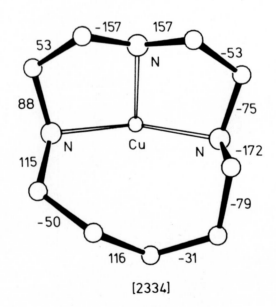

[2334]

Figure 3.7. Complexes of tridentate cyclodecanes [Mo([12]ane-
P$_3$)(CO)$_3$] (79) and [Cu([12]ane-N$_3$)]$^{2+}$ (78)

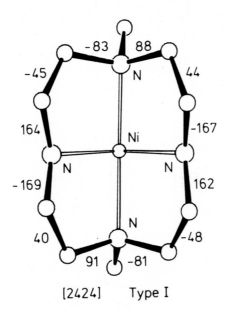

[2424]   Type I

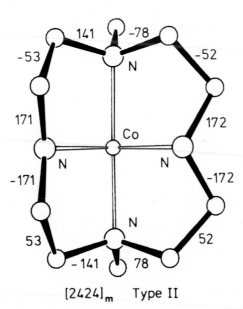

[2424]$_m$ Type II

Figure 3.8.   12-membered macrocycles in cis octahedral complexes
[Ni(Me$_2$-[12]ane-N$_4$)]$^{2+}$ (85) and [Co(Me$_2$-[12]ane-N$_4$)]$^{3+}$ (86)

An important aspect of macrocyclic stereochemistry not explicitly discussed before, is the effect of chiral atoms in the ring. Because of the rigidity imparted to the molecule by ring closure, this stereochemistry is by no means restricted to carbon atoms, but also includes the hetero donor atoms. Tsuboyama and co-workers (95-101) have studied the stereochemistry of a series of twelve-membered macrocyclics obtained by the tetramerization of substituted aziridines:

$$Et-\underset{\underset{CH_2}{\diagdown\diagup}}{\overset{\overset{R}{|}}{C}}\!\!-\!\!N-CH_2Ph \longrightarrow \left[CH_2-\underset{\underset{R}{|}}{\overset{\overset{Et}{|}}{C}}-\underset{\underset{CH_2Ph}{|}}{N} \right]_4$$

Only one isomer of the tetramer was found, which means that the ring cleavage occurs at the methylene-nitrogen bond. Starting from the racemic monomer, six different configurations are possible and these resolve into four geometrical isomers, including two racemic forms, on fractional crystallization (96):

Configurations

Geometrical Isomers

It is noted that, in general, the chirality of carbon in macrocycles requires two substituents, $R_1 \neq R_2$ and in addition a substituted macrocyclic nitrogen atom has chirality only if one of the neighbouring carbon atoms is substituted. The substituent on the nitrogen can be a chelated metal ion and nitrogen chirality therefore occurs in metallic macrocyclics of substituted ligands.

The most likely square [3333] conformation of the macrocycle occurs frequently, but not always with the same corner atoms which could be either CH_2, N or CHEt. The fundamental conformations for methylene corners are shown in Figure 3.9 and the possible distribution of corners in Figure 3.10. Where atoms of different chirality co-exist in the same twelve-membered ring, there is the further possibility that the corner atoms could be of different types to define rectangular or trapezoid conformations. This actually happens, as shown in Table 3, which summarizes the stereochemical details of twelve-membered chiral macrocyclics. The four distinct conformational types, [2334], [2343], [2424] and [3333] as found in the isomeric complexes of $CuCl_2$ and 1,4,7,10-tetrabenzyl-2,5,8,11-tetraethyl-1,4,7,10-tetraazacyclododecanes are illustrated in Figure 3.11. In this family of compounds there is a strict relationship between chirality, chelate ring conformation and macrocyclic conformation. This interesting correlation could be a significant general principle and should be further explored for other macrocyclics.

Thirteen-membered macrocyclic complexes are less plentiful. Only three have been characterized. The 1:1 Cu(II) complexes of [13]ane-S_4 and [13]ane-N_3O (32, 27) are both square pyramidal, with H_2O and Br$^-$ respectively as additional ligands, and with the macrocycle conformation [3334]. The Ni(II) low-spin complex of 3,3-Me$_2$-[13]ane-N_4 (102) has distorted square-planar coordination, and the same [3334] conformation. In all cases the corners occur at the central carbon of the propylene bridge, as shown in Figure 3.12. The macrocycle has type I structure in the Cu(II) complexes and type III in Ni(II). The disorder, so prevalent in fifteen-membered macrocyclics with opposed 5 and 6-membered chelate rings, is not observed here. The most likely explanation is the fact that the metal ion is too large to be in the plane of the donor atoms.

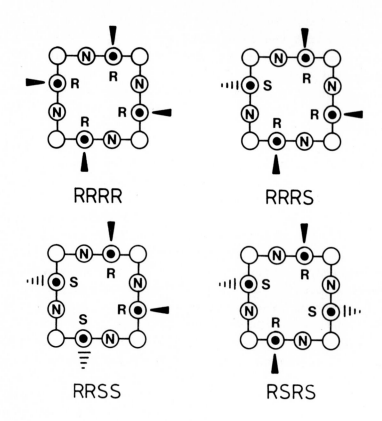

Figure 3.9. The relative stereochemistry of the four configurations of tetrameric aziridines with methylene corners in
[3333] macrocyclics

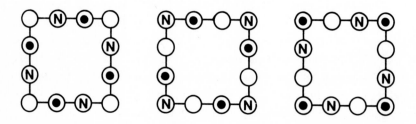

Figure 3.10. Three possible arrangements of the corner groups
from left to right; methylene corners, N corners and ethyl
corners. The asymmetric carbon is represented by ⊙

TABLE 3
Stereochemistry of 2,5,8,11-tetraethyl-1,4,7,10-tetraazacyclododecanes

N-substi-tuents	Cation	Coordination No.	Coordination Geometry	Macrocyclic Conf.	Macrocyclic Type	Chirality C	Chirality N	Ref.	Remarks
	Cu(II)	5	Sq.Pyramid	[3333]	I	RRRR		89	CH_2-corners
				[3333]	I	RRRR	RRRR	90	
Benzyl				[3333]	I	RRRR	RRRR	95	N-corners, square
Benzyl				[3333]	V	RSRS	RSRS	96	CH_2-corners, tub
Benzyl				[3333]	III	RRSS	RRSS	97	CH_2-corners, armchair
Benzyl				[3333]	II	RRRS	RRRS	98	CH_2-corners, square
Benzyl	Cu(II)	5	Dist. Sq. Pyramid	[2334]	I	RRRS	RRRR	99	λλλδ, C_1, (Figure 3.11)
Benzyl	Cu(II)	5	Sq.Pyramid	[2343]	I	RRSS	RRRR	100	λλδδ, C_i (Figure 3.11)
Benzyl	Cu(II)	5	Trigonal bipyramid	[2424]	I	RSRS	RRRR	101	λδλδ, S_4 (Figure 3.11)
Benzyl	Cu(II)	5	Sq.Pyramid	[3333]	I	RRRR	RRRR	89	λλλλ, C_4, Et-corners

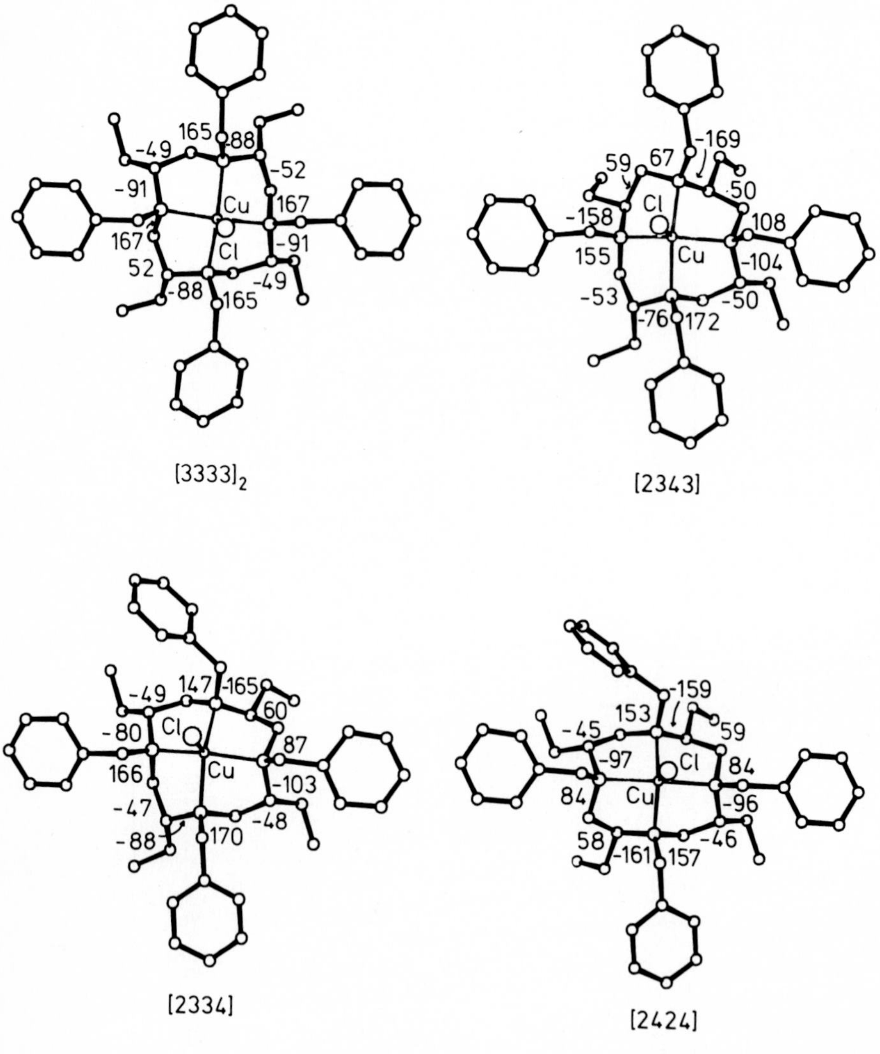

Figure 3.11. Four conformational types found in isomeric complexes of $CuCl_2$ and tetraethyl tetrabenzyl [12]ane-N_4

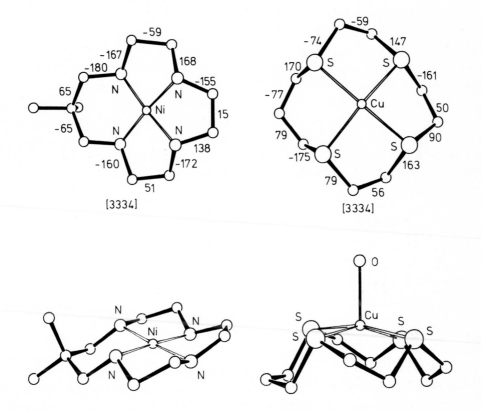

Figure 3.12. 13-membered macrocyclics; [Ni(12,12-Me$_2$-[13]ane-N$_4$)]$^{2+}$, macrocycle type III and [Cu([13]ane-S$_4$)]$^{2+}$, type I. Both have a [3334] conformation

3.3 LARGE MACROCYCLICS

The most extensive and important group of metallic macrocyclics includes compounds with fourteen to sixteen-membered rings, and of these the fourteen-membered family is the most populous. For the purpose of this review more extensive rings are classified as super large.

3.3.1 FOURTEEN-MEMBERED MACROCYCLICS

A fourteen-membered ring is sufficiently flexible to accommodate most preferred coordination geometries without serious steric consequences. Although the macrocyclic ring therefore readily assumes a planar configuration in many complexes, folded conformations are equally likely to occur. For macrocyclics of the $1,4,8,11-X_4$ type the conformations [124124] and [133133] are the most common, with the one-bond sides in propylene bridges.

For planar macrocycles the conformations [3434], [133133] and [3344] occur with the highest frequency. As shown below, the [133133] form is intermediate between the [3434] and [3344] conformations. It is noted that the one-bond sides now invariably occur in ethylene bridges.

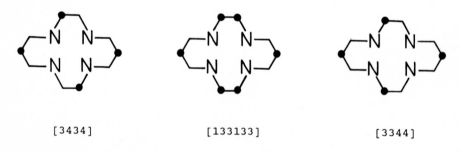

[3434]   [133133]   [3344]

The [3434] conformation occurs for complexes with a molecular centre of symmetry or a 2-fold axis perpendicular to the molecular plane. The [3344] conformation is compatible with a mirror plane bisecting the ring through the transposed propylenes or a 2-fold axis in the molecular plane. The intermediate form is compatible with all these possible symmetry elements, and is the only possible form for a complex at a site of 2/m crystallographic symmetry, Figure 3.13.

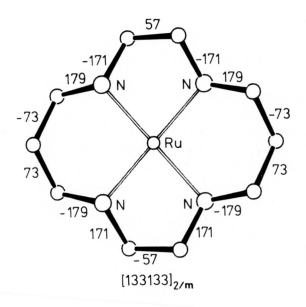

[133133]₂/m

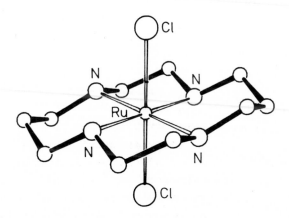

Figure 3.13. The complex [Ru([14]ane-N₄)Cl₂]⁺ on a site of 2/m symmetry showing [133133] conformation and coordination geometry

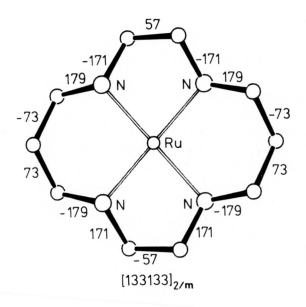

[133133]$_{2/m}$

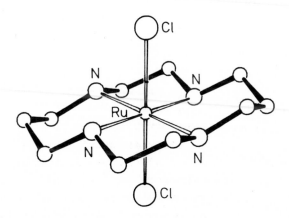

Figure 3.13.   The complex [Ru([14]ane-N$_4$)Cl$_2$]$^+$ on a site of 2/m symmetry showing [133133] conformation and coordination geometry

3.3.1.1   Cyclam, [14]ane-1,4,8,11-N$_4$

In the uncomplexed form, as the dihydroperchlorate, this unsubstituted macrocycle has the [3434] conformation. The molecule is centrosymmetric (103) with two intramolecular hydrogen bonds between adjacent nitrogen atoms, closing two six-membered rings.

Structural details of the crystallographically characterized complexes of cyclam are summarized in Table 4. In its complexes with Mo(II), Co(III) and Ru(III) the macrocycle has a folded [124124] conformation. In the Mo(II) and Ru(III) complexes the folding is a result of the metal ion size. In the Co(III) complex the coordination sphere is completed by ethylenediamine, which requires cis coordination and so forces the macrocycle to fold, Figure 3.14. When the coordination sphere is completed by moncdentate ligands, the Co(III) complex has trans configuration with the metal ion at the centre of a planar [133133] macrocycle (42). In the trans octahedral and tetragonally elongated complexes the macrocycle has either [133133] or [3434] conformation, invariably with a molecular centre of symmetry, Figure 3.14.

In the square planar Pd(II) complex the cyclam has a molecular mirror plane and the conformation [3344], whereas the centrosymmetric conformation [133133] occurs in the Pd(IV) complex (117). These two forms can also occur together in a mixed crystal, with an apparently uniform configuration of [133133] of the macrocycle in both square planar and octahedral environments. However, the mixed-stack arrangement is disordered with respect to the axial ligands, Cl$^-$. This can only occur if the alternating sequence in the stack is occasionally interrupted by consecutive square planar molecules. The crystallographic analysis therefore averages over [3344] and [133133] forms to yield the higher symmetry, interpreted as [133133] only (117). Direct evidence of the disorder is not apparent from the crystallographic analysis. Disorder of this type has also been observed a few years ago (120) in a trans Pt(II)-Pt(IV) bis(trans-1,2-cyclohexane) compound.

The small energy difference between the [3434] and [133133] forms is illustrated by their coexistence at independent centres of symmetry in the complexes of both Tc(V) (111) and Cr(III) (114). In the latter case the different conformations result from slightly different modes of hydrogen bonding with the

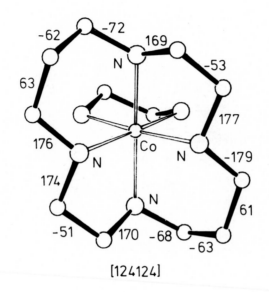

[124124]

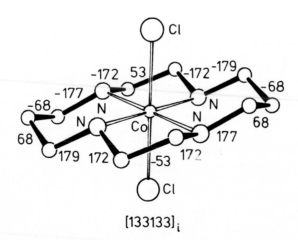

[133133]$_i$

Figure 3.14. Cis (107) and trans (42) Co(III) cyclam complexes

TABLE 4
Structural details of crystallographically characterized complexes of cyclam.
Coordination geometries are abbreviated by :
Oh = Octahedral, SP = Square Planar, TEO = Tetragonally distorted octahedron

Cation	Coordination		Macrocyclic		Ref.	Remarks
	Geometry	Symm.	Conf.	Type		
2(H⁺)		$\bar{1}$	[3434]	III	103	
Cu(II)	Oh	$\bar{1}$	[3434]	III	104	Trans $(ClO_4)^-$
Ni(II)	Oh	$\bar{1}$	[3434]?	III	105	Trans Cl^-; no coordinates published
Pb(II)	Oh		[124124]	V	106	Cis$(NO_3)^-$, pseudo bidentate
Co(III)	Oh		[124124]	V	107	Cis ethylene diamine, Cl^- anions
Ru(III)	Oh		[124124]	V	108	Cis Cl^-
Cr(III)	Oh		[124124]	V	109	Cis Cl^-, Cl^- anion
Co(III)	Oh	$\bar{1}$	[133133]	III	42	Trans Cl^-
Zn(II)	Oh	$\bar{1}$	[3434]	III	33	Trans$(O_2COCH_3)^-$ intermolecular bridging
Ru(III)	Oh	2/m	[133133]	III	110	Trans Cl^-, Br^- anion
Tc(V)	Oh	$\bar{1}$	[3434] [133133]	III III	111	2 Independent crystallographic units, Trans O
Ag(II)	SP	m	[3344]	I	112	Needle crystal, disordered Pbnm, Ag(II) 0.24 Å outside sq. coord. plane

Ag(II)	Oh	$\bar{1}$	[3434]	III	112	Block crystal, P$\bar{1}$, Ag(II) in macrocyclic coord. plane, Trans(ClO$_4$)$^-$
Ni(III)	Oh	$\bar{1}$	[3434]	III	113	Trans Cl$^-$
Cr(III)	Oh	$\bar{1}$	[3434], [133133]	III III	114	2 Independent crystallographic units, Trans carbamato, ClO$_4^-$ anion
Cu(II)	TEO	$\bar{1}$	[3434]	III	115	S-coordinated trans (SC$_6$F$_5$)$^-$
Cu(II)	TEO	$\bar{1}$	[133133]	III	116	Intermolecular trans (O$_2$COCH$_3$)$^-$ bridging
Pd(II/IV)	SP/Oh	$\bar{1}$	[133133]	III	117	Disordered mixed stack
Pd(II)	SP	m	[3344]	III	117	
Pd(IV)	Oh	$\bar{1}$	[133133]	III	117	Trans Cl$^-$
Ni(II)	Pseudo-Oh	$\bar{1}$	[3434]	III	118	Trans bridging I$^-$
Co(II)[1]	TEO	$\bar{1}$	[133133]	III	119	Trans ClO$_4^-$

uncoordinated oxygen atoms of the secondary carbamato ligands. A similar effect probably operates in the Tc-complex.

3.3.1.1.1   C-substituted cyclams

Just as the direction of lone pairs and hydrogen positions can decisively influence the stereochemistry of the small macro-cyclics, substituents on both donor and carbon atoms in the ring, could be sterically important for large macrocyclics. This field has not been explored in any detail beyond the introduction of methyl substituents or pendant donor groups. As noted before, modifications of this type further complicates the stereochemistry by the introduction of chiral centres.

The simplest C-substituted derivative of cyclam that has been characterized is $5,12-Me_2-[14]ane-N_4$. As a free ligand it has a [3434] conformation with 2-fold symmetry (121). The [3434] folding is presented in the centrosymmetric complexes with Ni(II) (122) and Co(III) (123). Two tetramethyl derivatives have been structurally characterized as Ni(II) complexes. The $5,6,12,13-Me_4$ cyclam has a [3434] conformation (124), whereas $5,7,12,14-Me_4$ cyclam has a [133133] conformation in the low-spin (125) and [3434] in the high-spin complex. Structural details of these complexes are summarized in Table 5.

By far the most widely studied are the complexes of hexasubstituted $5,5,7,12,12,14-Me_6$ cyclam. The meso isomer (Figure 3.15) is often called _tet a_ while the racemic isomer (Figure 3.15) is called _tet b_.

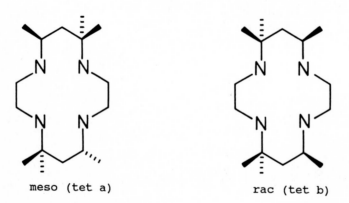

meso (tet a)   rac (tet b)

Figure 3.15.   The meso and rac isomers of 5,5,7,12,12,14-Me6 cyclam

A further subdivision into α,ß and γ forms is often introduced to distinguish between ligand types V, I and III respectively. Stereochemical details of crystallographically characterized metal complexes of Me$_6$-cyclams are summarized in Table 6.

There is only one α-tet a (Type V) structure reported in the literature (139). It is the cis Ni(II) macrocyclic with the irregular [12434] conformation, Figure 3.16. Stability studies (152) showed that α-tet a forms complexes less stable than those of α-tet b. This could be due to the inability of α-tet a to bind the metal ion symmetrically. The Ni(α-tet b) macrocyclic (147) has a [133133] conformation, Figure 3.16. Macrocyclics of γ-tet a (type III) have been studied more often and a [3434] conformation has been observed for square-planar and octahedral complexes, and a [133133] conformation for trans octahedral macrocyclics.

In the tet b series the α-form (type V) complexes metal ions with [3434] conformation in the square-planar Ni(II) low-spin macrocyclics and [133133] conformation in trigonal bipyramidal Cu(II) complexes. The macrocycle also binds to a variety of metal ions with a cis octahedral geometry and in these complexes it has either a [124124] or [133133] conformation. Only one structure each has been reported for ß-tet b and γ-tet b. Both are of low-spin Ni(II) complexes. The ß-tet b form has a [3344] conformation and there is slight tetrahedral distortion around the metal ion (146). The γ-tet b form (146) has a [3434] conformation.

3.3.1.1.2   N-substituted cyclams

Complexes, only of the ligand 1,4,8,11-tetramethyl-(1,4,8,11-Tetraazacyclotetradecane), tetramethylcyclam or TMC, have been studied in some detail. The results are summarized in Table 7. Possible stereoisomeric forms of the macrocycle are (157) as follows:

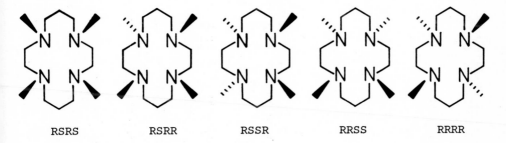

| RSRS | RSRR | RSSR | RRSS | RRRR |

TABLE 5
Structural details of crystallographically characterized complexes of C-substituted cyclams, excluding Me$_6$ derivatives

Substituents	Cation	Coordination		Macrocyclic		Relative Stereochemistry		Ref.	Remarks
		Geometry	Sym.	Conf.	Type	C	N		
5,12,-Me$_2$			2	[3434]		SS		121	
meso 5,12-Me$_2$	Co(III)	Oh	$\bar{1}$	[3434]	III	RS	RSSR	123	Trans(N$_3^-$)$_2$, N$_3^-$ anion
5,2-Me$_2$	Ni		$\bar{1}$	[3434]	III	RS	RSSR	122	
5,6,12,13-Me$_4$	Ni(II)		$\bar{1}$	[3434]				124	
5,7,12,14-Me$_4$	Ni(II)[l]	SP	$\bar{1}$	[133133]	III	RSSR	RSSR	125	ClO$_4^-$ anions
5,7,12,14-Me$_4$	Ni(II)[h]	Oh	$\bar{1}$	[3434]	III	RSSR	RSSR	125	Trans(ClO$^-$)$_2$
5,12-Me$_2$-7,14 Pr$_2^i$	Ni(II)		$\bar{1}$	[133133]	III			126	Pink form, ClO$_4^-$
5,12-Me$_2$-7,14-Pr$_2^i$	Ni(II)		$\bar{1}$	[3434]	III			126	violet form, ClO$_4^-$
2,5,5,7,9,12,12,14-Me$_8$	Cu(II)	SP	$\bar{1}$	[3434]	III	SRRS	SRRS	127	(ClO$_4^-$)$_2$ anions

5,7,12,14-Et$_4$-7,14-Me$_2$	Cu(II)	SP/tetr.	$\bar{1}$	[3434]	III	SRRS	RRSS	128	Axial (ClO$_4^-$)$_2$
5,12-7,14-Me$_2$Ph$_2$	Cu(II)					RRSS	SSRR		No coordinates
2,5,5,7,9,12,12,14-Me$_8$	Ni(II)[1]	Td.dist.	2	[3434]	I	RSRS	RSRS	129	(ClO$_4^-$)$_2$ anions
5,7,12,14-Et$_4$-7,14-Me$_2$	Ni(III)		$\bar{1}$	[3434]			RSSR	130	(ClO$_4^-$)$_2$

TABLE 6
Stereochemical details of meso Me$_6$ cyclam (teta) and rac Me$_6$-cyclam (tetb) metal complexes

Isomer	Cation	Coordination		Macrocyclic		Relative stereochemistry		Ref.	Remarks
		Geometry	Sym.	Conf.	Type	C	N		
tetb	H$_2$O			[133133]				131	
α-tetb	Co(II)	Oh		[124124]	V	SS	RRRR	132	Cis bidentate R-
γ-teta	Ni(III)	Oh	$\overline{1}$	[3434], [133133]	III	RS	RRSS	133	Two independent units Trans (H$_2$PO$_4^-$)$_2$, ClO$_4^-$ anion
α-teta	Ni(II)	Oh		[124124]	V	SS	RRRR	134	Dimer, linked by cis bidentate hydrated d- tartrate
α-teta	Ni(II)	Oh		[133133]	V	SS	RRRR	135	Cis (H$_2$O)$_2$, Cl$^-$ anions
α-tetb	Hg(II)	Oh	2	[133133]	V	SS	RRRR	136	Cis (μCl$_2$)-Hg Cl$_2$
α-tetb	Cu(II)	Trigonal bipyramid		[133133]	V	SS	RRRR	137	Equatorial NO$_3^-$, ClO$_4^-$ anion
γ-teta	Ni(II)	Oh	$\overline{1}$	[3434]	III	RS	RSSR	138	Trans (F$^-$)$_2$, 5H$_2$O
α-teta	Ni(II)	Oh		[12434]	V	RS	RRRR	139	Cis Acac
α-tetb	Cu(II)	Trigonal bipyramid		[133133]	V	SS	RRRR	140	Equatorial o-mercapto-benzoate
γ-teta	2(H$_2$O)			[3434]	III	RS		141	-N---H-O$^-$H--H-N- bridging
γ-teta	Cu(II)	Oh		[3434]	III	RS	RSSR	142	Trans (ClO$_4^-$)$_2$

γ-teta	Ni(II)[l]	SP	$\bar{1}$	[3434]	III	RS	RSSR	143	Trans $(Cl^-)_2$, $2CHCl_3$
γ-teta	Ni(II)[h]	Oh	$\bar{1}$	[3434]	III	RS	RSSR	143	Br^- anions, $2H_2O$
γ-teta	Ni(II)[l]	SP	$\bar{1}$	[3434]	III	RS	RSSR	144	Red form, Trans $(ClO_4^-)_2$
γ-teta	Cu(II)	Oh	$\bar{1}$	[3434]	III	RS	RSSR	145	ClO_4^- anions
α-tetb	Ni(II)[l]	Dist. SP	2	[3434]	V	SS	RRRR	146	
β-tetb	Ni(II)[l]	Dist. SP		[3344]	I	SS	SRSR	146	$ZnCl_4^{2-}$ anion, H_2O
γ-tetb	Ni(II)[l]	SP		[3434]	III	SS	SSRR	146	ClO_4^- anions
α-tetb	Ni(II)[h]	Oh	$\bar{1}$	[133133]	V	SS	RRRR	147	Cis bidentate $CH_3-C(=O)(-O^-)$, ClO_4^-
γ-teta	Ag(II)	SP	$\bar{1}$	[3434]	III	RS	RSSR	148	NO_3^- anions, $d(Ag-O)=2.8$Å
γ-teta	Cr(III)	Oh		[3434]	III	RS	RSSR	149	Trans Cl^-, H_2O; NO_3^- anions
α-tetb	Cr(III)	Oh	2	[124124]	V	SS	RRRR	150	Cis $(OH^-)_2$
α-tetb	Cr(III)	Oh		[1213133]	V	SS	RRRR	150	Cis (O_2CO^{2-})
α-tetb	Cu(II)	Trigonal bipyramid	2	[133133]	V	SS	RRRR	151	Dimer with Cl^- bridge

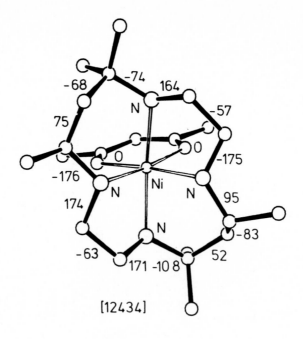

[12434]

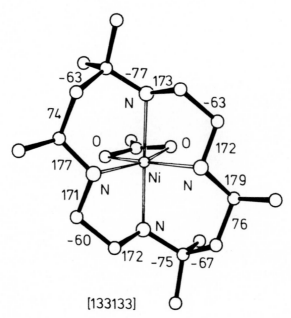

[133133]

Figure 3.16. Structural comparison of the [12434] αteta (139) and [133133] αtetb (147) complexes of Ni(II)

TABLE 7
Stereochemical details of crystallographically characterized metal complexes of N-substituted tetramethylcyclam

Isomer	Coordination		Macrocyclic		N Stereo-chemistry	Ref.	Remarks
	Geometry	Symm.	Conf.	Type			
Ni(II)[h]	Oh	$\bar{1}$ dimer	[3344]	III	RSSR	153	Trans(N_3^-)$_2$, bridging N_3^-, L^-
Zn(II)	Sq.Pyramid		[3344]	I	RSRS	154	Apical Cl^-, Zn 0.57 Å above N_4
Ni(II)	Sq.Pyramid	m	[3344]	I	RSRS	155	Apical Cl^-, ClO_4^-, Ni 0.33 Å above N_4
Fe(II)	Sq.Pyramid		[3434]	I	RSRS	156	Apical NO^-, N_4 not planar
Fe(III)	Oh	m	[3434]	I	RSRS	156	Trans(NO^-,OH^-), Fe 0.15 Å above N_4
Ni(II)[l]	SP	2	[3434]	I	RSRS	157	($CF_3SO_3^-$) anions, H_2O, acetone
Ni(II)[l]	SP	2	[3434]	I	RSRS	158	(ClO_4^-) anions, disorder
Ni(II)[h]	TBP/Sq. Pyramid	2	[3434]	I	RSRS	116	Equatorial (O_2COCH_3), in Trigonal bipyramidal description
Ni(II)[h]	Oh	m	[3344]	III	RSSR	116	Trans(O_2COCH_3)$_2$
Ru(IV)	Oh		[3434]	II	RSRR	159	Trans(Cl^-,O), ClO_4^-
Zn(II)	TBP		[3434]	I	RSRS	33	Equatorial O_2COCH_3

The stereochemistry refers to the four chiral nitrogen atoms, noting the need to distinguish between types III and IV.

Most of the macrocyclics in this class, investigated to date are of type I. This means that all methyl groups are on the same side of the coordination plane, shown to be the energetically most favourable arrangement, according to molecular mechanics calculation (157). In most cases, the metal ion is also displaced from the coordination plane, invariably in the same direction as the methyl groups. An investigation with space-filling models revealed (156) that coplanarity of the donors and metal causes the methyl groups to occupy positions very close to a possible axial ligand. Lifting the metal ion out of the plane allows the nitrogen atoms to bend back, thus relieving the steric crowding, while maintaining a planar array of donors, Figure 3.17. Coordination number for type I TMC rarely exceeds five. The Fe(III) complex with a distorted octahedral arrangement (156) is the only known exception.

If the macrocycle has a type III skeleton trans octahedral coordination of metals is more feasible as in the high-spin Ni(II) complexes (153, 116). Ru(IV) in the type II macrocyclic (159) also has trans octahedral coordination. In all cases however, the macrocycle has either [3344] or [3434] conformation, irrespective of metal-ion geometry, Figure 3.17.

Barefield and co-workers (160) have prepared cyclams with 2-cyanoethyl, 2-carbamoylethyl, or a combination of these as N-substituents. Tetracyanoethylcyclam (161) coordinates octahedrally with high-spin Ni(II), in either type III or type I (161) configuration and a [3434] conformation. Tetra(2-carbamoylethyl)-cyclam is type III with a [3344] conformation and with two coordinating pendant groups in its trans octahedral high-spin Ni(II) complex (161). The 1,8 dicarbamoylethyl-4-cyanoethyl derivative (161) is folded [124124] with the carbamoylether groups coordinating.

Murase and co-workers prepared (162, 163) tetra(2-aminoethyl)cyclam which is able to coordinate two metal ions with its eight donor atoms, in addition to the formation of mononuclear complexes, with non-coordinating pendant groups. The macrocyclic ligand of structure

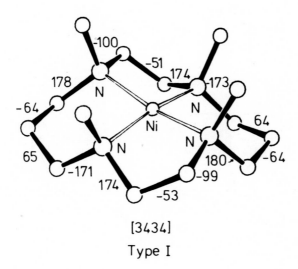

[3434]

Type I

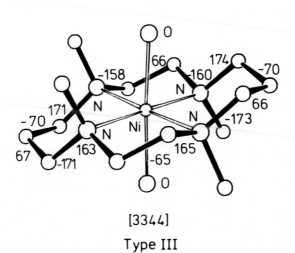

[3344]

Type III

Figure 3.17. Comparison of the $[Ni(Me_4-[14]ane-N_4)]^{2+}$ complexes of type I (154) and type III (116) respectively

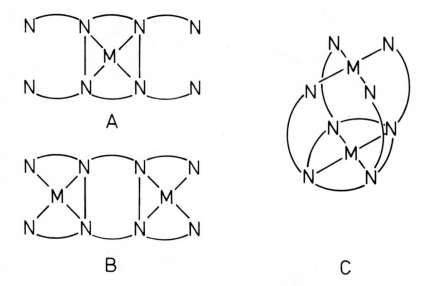

was postulated to form metal complexes of the types

Structures of two Cu(II) and a low-spin Ni(II) complex have
been studied (163) and found to have coordination structures of
the B-type. In fact, two different arrangements of this type were
found:

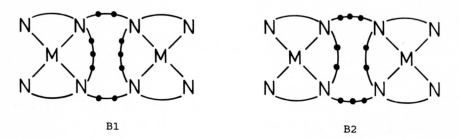

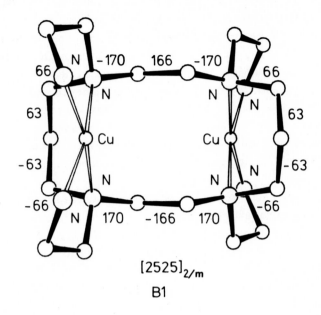

[2525]$_{2/m}$

B1

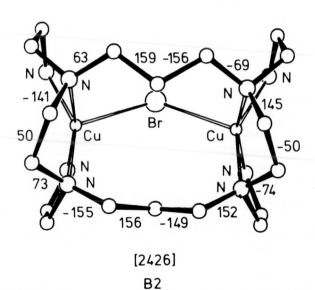

[2426]

B2

Figure 3.18. Cu(II) complexes of tetraamino-ethylcyclam showing two different coordination modes [Cu$_2$(taec)]$^{4+}$ (162) and [Cu$_2$(taec)Br]$^{3+}$ (163)

The nickel and one of the copper complexes have B1 coordination with the macrocyclic in [2525] conformation. The nickel coordination is square planar and the copper is square pyramidal with an axially coordinated ClO_4^- anion. In the second copper complex (163) there is a bridging bromide ion between the two copper ions, in a B2, trigonal bipyramidal coordination mode. The macrocyclic conformation is [2426], Figure 3.18.

A similar ligand, cyclam-1,4,8,11-tetraacetic acid also forms a binuclear complex with Cu(II) (93). The metal ion is square pyramidal with 2 amine and two carboxylic donors in the plane and either a water molecule or a free carboxylate donor at the apex. The coordination mode is B1 and the macrocycle conformation is [2525].

An interesting new fourteen-membered tetraazamacrocyclic ligand 5,5,7,12,12,14-hexamethyl-1,4,8,11-cyclam-N',N'''-diacetic acid (164) coordinates high-spin Ni(II) octahedrally, with the pendant donors in cis relationship and the macrocycle in [12434] folded conformation.

The only example of a dimeric cyclam is the Ni(II) complex that forms during nickel-assisted cyclization of linear tetraamines in macrocyclic synthesis (165). As shown in Figure 3.19, the two halves of the dimer are related by a crystallographic 2-fold axis. The nickel is in a square-planar environment in the type III cyclam moiety with a [3344] conformation.

3.3.1.2   Isocyclam, [14]ane-1,4,7,11-N$_4$

Isocyclam is the isomeric modification of cyclam obtained by interchanging an ethylene and a propylene bridging group:

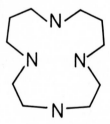

The only simple complex of isocyclam that has been characterized crystallographically is the disordered low-spin square planar

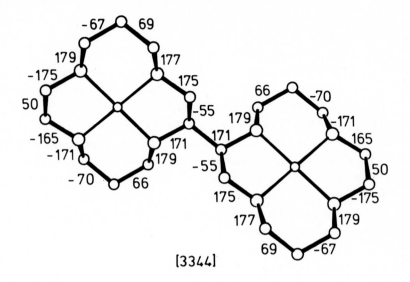

[3344]

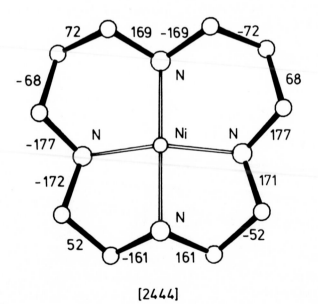

[2444]

Figure 3.19. Structure of the dimeric cyclam, [Ni(cyclam)]$_2^{4+}$, (top) and the low spin [Ni(isocyclam)]$^{2+}$ structure, (bottom), (165, 28)

complex of Ni(II) (28). The macrocycle has a type II arrangement with the unique hydrogen atom at the junction of the five-membered rings. The molecule has a mirror plane and a [2444] macrocyclic conformation, Figure 3.19.

The same type II structure with a [2444] conformation was observed (166) in the low-spin Ni(II) complexes of ethylene- and xylylene-bridged dimeric cyclams, bridged at the N-junction between the six-membered chelate rings.

An isocyclam with a 2-dimethylaminoethyl pendant ligand at the bridgehead between six-membered rings, formed isomorphous high-spin Ni(II) and Cu(II) complexes (167). Coordination of the pendant group results in a trigonal bipyramidal complex. The macrocyclic arrangement is somewhat doubtful, because of disorder, but most likely represents a different type II configuration, with the unique hydrogen at a junction between five- and six-membered rings. To accommodate the pendant arm an additional corner is formed and the ring assumes a [13424] conformation.

3.3.1.3   [14]Ane-1,4,7,10-N$_4$

In this isomeric modification of cyclam three of the nitrogen atoms are connected by ethylene groups, with a butylene link closing the macrocycle. When coordinating as a tetradentate ligand a 5,5,5,7 sequence of chelate rings is therefore established.

It forms both trans (168) and cis (169) high-spin octahedral complexes, with NiCl$_2$ and Ni(NCS)$_2$ respectively. In the cis complex the macrocycle is of type III and adopts a [13325] conformation whereas the trans arrangement is formed by type V, adopting a [14234] conformation, Figure 3.20.

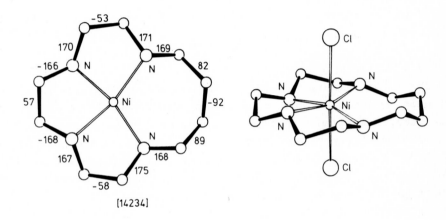

[14234]

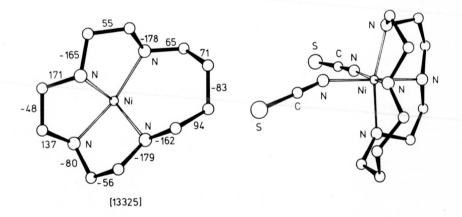

[13325]

Figure 3.20. Complexes of trans (type V) and cis (type II) [Ni(1,4,7,10-[14]ane-N$_4$)]$^{2+}$, (168, 169)

The unit cell of the $Cu(ClO_4)_2$ complex contains two independent monomeric units (170), one with 2-fold symmetry, possibly due to disorder. The observed [124142] folding observed for the latter could be due to the disorder, with the actual molecular geometry represented by the other moiety of [14234] conformation. The perchlorate groups are coordinated in trans relationship and, as for the nickel complex, the macrocycle has type V structure.

It is noted that the seven-membered chelate ring introduces at least one extra corner, with a minimum of five sides, compared to four of the regular [14]anes.

3.3.1.4   S-donor macrocyclics

The sulfur analogue of cyclam and its metal complexes have been studied in considerable detail, as summarized in Table 8. As expected, one finds that the second-period heteroatom imparts extra flexibility to the macrocyclic ring, manifested here by the ability of sulfur-donor macrocycles to form exodentate complexes, in addition to the more familiar endodentate variety. The coordination mode depends on the orientation of the lone pairs on the sulfur atoms, relative to the macrocyclic cavity. Specification of structure type in Table 8 also refers to the orientation of the sulfur lone pairs.

The uncoordinated ligand crystallizes in the form of an α and a β polymorph (171). In both crystals the molecule is centro-symmetric, with the sulfur atoms at the corners of a [3434] arrangement, and the lone pairs pointing out of the cavity, Figure 3.21. The β-polymorph contains two independant molecules, in one of which the ethylene bridges have a sterically unlikely configuration, no doubt an artifact of the observed disorder. Disorder of this type is also associated with the same unlikely structure of the macrocycle where it acts as a double monodentate intermolecular bridge between two $NbCl_5$ units (178).

The relationship between the exodentate and endodentate forms of the ligand was discussed by De Simone and Glick (171). Contrasting the exodentate [3434] with the endodentate [3434] arrangement, found in the Cu(II) and Ni(II) complexes, the two forms seem to be related by pseudo-rotation, in which each anti interaction is rotated into a gauche interaction and vice versa.

TABLE 8
Stereochemistry of [14]ane-S_4 macrocyclics

Cation	Coordination			Macrocyclic		Ref.	Remarks
	Geometry	Symm.	Mode	Conf.	Type		
None		$\bar{1}$	exo	[3434]		171	S at corners. α-form
None		$\bar{1}$	exo	[3434]	I	171	S at corners. β-form
Hg(II)	Sq.Pyramid		endo	[13343]	I	172	Apical H_2O, ClO_4^- anions; Hg 0.48 Å above S_4
Hg(II)	Td	$\bar{1}$	exo	[124124]		172	$(HgCl_2)_2L$, S at corners
Hg(II)	Td	$\bar{1}$	exo	[3434]		173	$(HgI_2)L$, S at corners
Ru(II)	Oh	2	endo	[124124]	V	174	Cis $(Cl^-)_2$; $2H_2O$
Cu(II)	Tetr. dist.Oh	1	endo	[3434]	III	175	Axial $(ClO_4^-)_2$, polymeric
Cu(I)	Dist.Td		3 endo	[223223]		176	Intermolecular exo coordination bond
Ni(II)	SP	$\bar{1}$	endo	[3434]	III	177	BF_4^- anions
Nb(V)	Oh	$\bar{1}$	exo	[3434]		178	Unidentate macrocycle, 2S bridge between Nb; S at corners
Rh(I)	SP		endo	[3344]	I	179	Van der Waals dimer

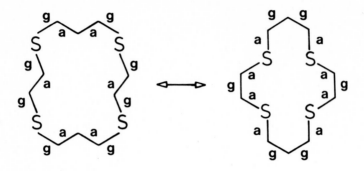

In 4n ring systems this would simply correspond to moving the sulfur atoms around a ring of approximately constant shape, but in 4n+2 rings this completely alters the shape of the ring. There are six gauche and eight anti interactions in the endocyclic conformation, with the inverse ratio for exodentate. Torsional strain would therefore favor the endodentate conformer, but clearly the mutual repulsion between sulfur lone pairs is sufficient to overcome this advantage and forces the exo conformation on the free ligand. A metal acceptor of lone-pair density into the cavity must stabilize the endodentate conformation. It is interesting to note that in mixed oxygen-sulfur polyethers with ring sizes of 12, 15 and 18, the sulfur atoms also have their lone-pairs directed outwards (180). It is clear however, that the energy difference between endo and exo conformations cannot be substantial.

When the ligand binds to two $HgCl_2$ molecules the sulfur atoms are also at the corners of a [124124] configuration and two of these complete each Hg(II) coordination tetrahedron, as shown in Figure 3.21. Complexing with the more ionic mercury (II) perchlorate, produces a square pyramidal endodentate complex with the mercury ion above the coordination plane, capped by an apical water molecule (172). The ligand is also endodentate in the $RuCl_2$ complex, but since the metal ion is too large for the cavity, the macrocycle folds into a [124124] conformation to form a cis-octahedral complex. This structure is shown in Figure 3.22. In the Ni(II) and Cu(II) complexes where the metals are small enough for encirclement, the sulfur atom are endodentate and the macrocycle has a regular [3434] conformation.

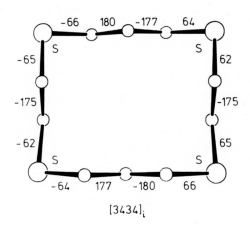

[3434]$_i$

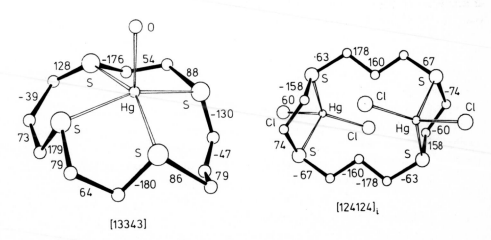

[13343]

[124124]$_i$

Figure 3.21. The conformation of [14]ane-S$_4$ as free ligand (171) and in the mercury complexes [Hg[14]ane-S$_4$)H$_2$O]$^{2+}$ and [(HgCl$_2$)$_2$([14]ane-S$_4$)] (172)

The Rh(I) complex of [14]ane-S_4 has a [3344] conformation. The geometry around the metal is square planar, but since the ligand is type I rather than type III short Rh---Rh and Rh---S intermolecular non-bonded interactions are observed, Figure 3.22. The tetrahedral requirements of Cu(I) are met (176) by three endodentate donors of one macrocycle and a fourth exodentate coordinating donor from a second macrocycle. The macrocyclic conformation appears to be [223223].

3.3.1.5   P-Donor macrocyclics

Metal complexes of only one authentic phosphorus-donor macrocycle, viz. 5,7,12,14-(OH)$_4$-1,4,5,7,8,11,12-Me$_8$-[14]ane-1,4,8,11-P$_4$, an analogue of cyclam, have been crystallographically characterized (181). Two isomeric Pd(II) complexes of the ligand were obtained, with macrocyclic structures of type III and type I respectively. Only in the former type is square-planar coordination achieved with a [133133] conformation and a 2-fold axis in the macrocyclic plane. However, there is evidence of disorder and the stereochemistry is not unassailable. In the type I isomer square-planarity cannot be attained, despite severe deformation of the central C-C-C angles of the propylene links. The Pd lies 0.26Å above the coordination plane. This activates a fifth coordination site, occupied by a Cl⁻ ion to complete the square-pyramidal coordination with the macrocycle in [3344] conformation.

The only other macrocyclic ligand with phosphorus donor atoms is actually a cyclam derivative with pendant phosphorus donor groups, 1,4,8,11-tetrakis(methyldiphenylphosphino)cyclam. The free ligand (182) has the rectangular [3434] conformation. The pendant group are in equatorial rather than axial positions and the propylene corners are not at the central carbon atoms. No metallic macrocyclics of the ligand have been reported yet.

3.3.2   Fifteen-membered macrocyclics

The stereochemistry of cyclopentadecanes is complicated by the high coincidence of disorder and too few structures are available to enable a systematic discussion of conformational trends. The known complexes fall into the classes of tetradentate and pentadentate macrocyclics. Tetradentate macrocyclics with 1,4,8,12-donor positions are generally disordered because of a

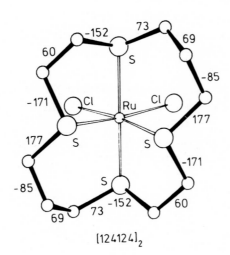

[124124]₂

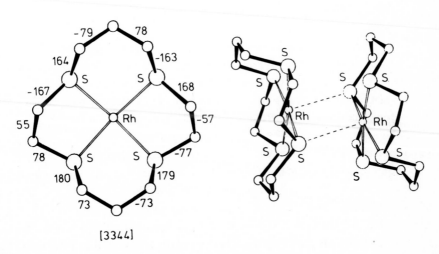

[3344]

Figure 3.22. The conformation of [14]ane-S₄ in cis [Ru([14]ane-S₄)Cl₂] (174) and dimeric [Rh[14]ane-S₄]⁺ (179). In the dimer the dashed lines indicate some of the short non-bonded distances

statistical superposition of the single five-membered chelate ring onto one of the three six-membered rings to yield metal complexes with apparent molecular centres of symmetry. The macrocycles are usually planar and the conformation appears to be [3444] in all cases (30-34). The 1,4,7,10-[15]ane-N_4 ligand binds to Ni(II) to form three 5-membered and one 8-membered chelate ring with an overall conformation of [123432] (168).

In the class of pentadentate macrocyclics each new structure represents a new conformational type, which is a measure of the ring flexibility. The free [15]ane-O_3S_2 has a [13443] conformation, with only two of the oxygen donors in endodentate positions (183). No metallic macrocyclics of this ligand have been structurally characterized.

A variety of isomers of the ligand [15]ane-N_2OS_2 can exist. Two of these,

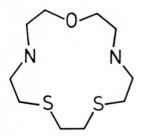 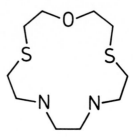

have been studied crystallographically. Ag(I) binds to both to form 5-coordinate square-pyramidal complexes, not involving oxygen atoms and with SCN^- in the apical position (184, 185). Both ligands are pentadentate in their Ni(II) complexes. In the former complex (186) a distorted octahedron is completed by a water molecule, trans to S and in the latter (187) a monodentate nitrato group trans to O has this function. The sulfur atoms are cis in both cases and the macrocyclic conformations are [123423] and [23424] respectively, Figure 3.23. A Pd(II) complex of the first ligand has been studied. It has 2 N and 2 S donors in a plane, with Pd above the plane in strong Pd-O interaction, d(Pd-O) = 2.78Å, to define square pyramidal coordination with the macrocycle in [1231323] conformation (188).

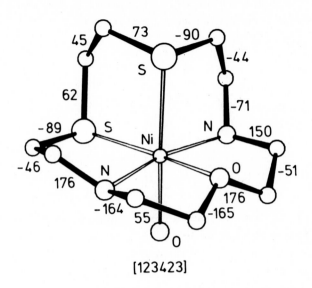

[123423]

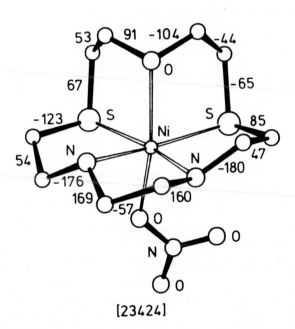

[23424]

Figure 3.23. N_2OS_2 ligands with two different donor arrangements which preserve a cis relationship of sulfur in their nickel complexes (186, 187)

3.3.3 SIXTEEN-MEMBERED MACROCYCLICS

Most of the known macrocyclic complexes in this class are based on the symmetrical tetradentate type of ligand,

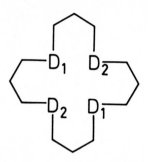

where D_1 and D_2 are either N or S donors. Of these, the most common are tetraaza compounds, $D_1 = D_2 = N$, summarized in Table 9. The free ligand, 1,5,9,13-[16]ane-N_4 has $\overline{4}$ symmetry and type V proton distribution (189). The conformation of this structure, shown in Figure 3.24 is [4444]. It assumes type I configuration in its complexes with Cd(II), Hg(II) and Pb(II), (190), but the conformation is still of the [4444] variety. The coordination number in these complexes hinges on a decision about the coordination mode of the nitrato groups, with probable values of seven for Cd(NO_3)$_2$ and up to eight for Pb(NO_3)$_2$, as shown in Figure 3.24. The HgCl$_2$ complex has a clearly defined trans octahedral coordination but the chloro group in the PbCl$_2$ is disordered between two equivalent capping positions for square pyramidal coordination. The coordination geometry of the Cd(NO_3)$_2$ complex of the tetramethylated derivative of the ligand is equally uncertain (191), and the situation is further complicated by unspecified disorder in the structure. Each of the three independent sites appears to be randomly occupied by macrocyclics in different modifications, presumably types I and II, although this has not been clearly established. Geometrical arguments that favour square pyramidal coordination are therefore not totally convincing. Distorted octahedral arrangements are more in line with apparent coordination numbers. Conformational assignments are equally tentative. The Ru(VI) complex of the same ligand has a [4444] conformation with the metal ion in the N_4 plane of the centrosymmetric complex (34).

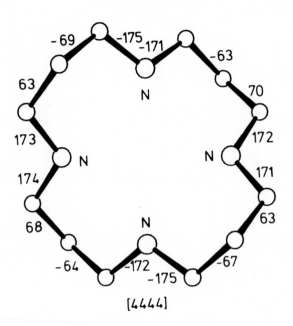

[4444]

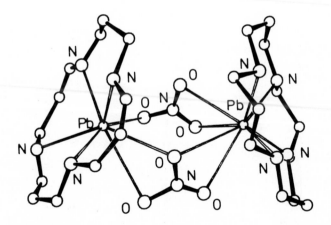

Figure 3.24. [16]ane-N_4 as a free molecule (189) and its lead complex, $[Pb([16]ane-N_4)]_2(NO_3)_2{}^{2+}$ (190)

TABLE 9
Stereochemistry of the macrocyclics of [16]ane-X_4 type, characterized by crystallography

Cation	Substituents	Coordination		Macrocyclic		Ref.	Remarks
		Geom.	Symm.	Conf.	Type		
	$D_1=D_2=N$						
Cd(II)		7-coord.	4	[4444]	V	189	Bidentate NO_3^-
Hg(II)		Oh	4mm	[4444]	I	190	Trans $2Cl^-$, bridging
Pb(II)		Sq.Pyr.	2	[4444]	I	190	Coord. Cl^- disordered
Pb(II)		Variable		[4444]	I	190	NO_3^- coordinated in various modes
Cd(II)	$1,5,9,13\text{-}Me_4$	Dist.Oh		[134314], [134134]	I/II, I	191	3 Independent units, some unspecified disorder, Bidentate NO_3^-, $[Cd(NO_3)_4]^{2-}$ anion
Ru(VI)	$1,5,9,13\text{-}Me_4$ $D_1=D_2=S$	Oh	$\bar{1}$	[4444]	III	34	Trans $2(O^{2-})$, ClO_4^- anion
Cu(II)		Oh	$\bar{1}$	[4444]	III	32	Trans $(ClO_4^-)_2$
Hg(II)		7-coord?		[4444]	V	192	Trans $(ClO_4^-)_2$, one bidentate, disordered
Mo(II)		Oh	dimer, $\bar{1}$	[4444]	I	193	Trans dimeric, Mo above plane, $d(Mo-Mo)^1 = 2.82$ Å
Mo(IV)		Oh		[4444]	I	194	Dimer, O_2^- and $CH_3CH_2O^-$ trans, (μ-oxo) bridge

Mo(IV)	D$_1$=N, D$_2$=S			195		Trans SH$^-$, O^{2-}, Mo above plane
Pd(II)	3,3,7,7,	Dist.SP	[134134]	196	III	2 PF$_6^-$ anions, H$_2$O
Pd(II)	11,11,15,15,-	Dist.SP m	[134314], [13444]	197	I, II	2 Cl$^-$ anions, 2H$_2$O
Ag(I)	Me$_8$	Sq.Pyr.	[4444]	198	V	Apical acetate

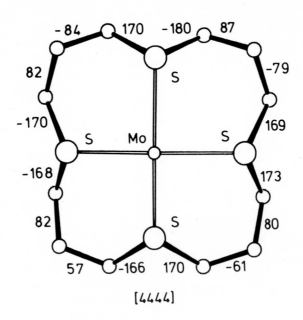

[4444]

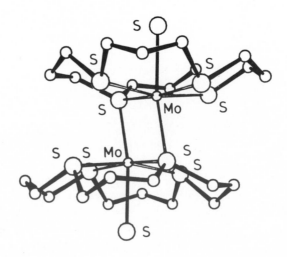

Figure 3.25. The conformation of [16]ane-S$_4$ in the molybdenum complex [Mo([16]ane-S$_4$)]$_2{}^{2+}$ (193) and the coordination geometry of the complex

Structural details of [16]ane-S$_4$ complexes (D$_1$ = D$_2$ = S) are also summarized in Table 9. Copper, mercury and molybdenum complexes have been studied. In all cases the metal ions have trans octahedral geometry and the macrocycle has a [4444] conformation. In the centrosymmetric Cu(II) complex the macrocycle is type III, in the Hg(II) it is type V and in the Mo(II) and Mo(IV) complexes the macrocycle is type I.

The coordination geometry of the Hg(II) complex is not defined unequivocally since the coordinated perchlorato group is disordered to such an extent that monodentate and bidentate arrangements thereof are equally likely, and seven-coordination is not rigorously excluded. The dimeric structure of the molybdenum (II) trifluoromethanesulfonate complex (193) is shown in Figure 3.25. There are two one-atom bridges and the distance between the molybdenum atoms of 2.82Å does not preclude an intermetal bond. Each Mo atom is displaced by 0.2Å from the coordination plane, towards the centre of symmetry.

Three complexes of an octamethylated mixed-ligand (D$_1$ = N, D$_2$ = S)[16]ane are listed in Table 9. CH$_3$COOAg forms (198) a square pyramidal complex with a type V macrocycle in [4444] conformation. The other structures are both of Pd(II) complexes. With a hexafluorophosphate counter-ion the macrocyclic has type III structure with Pd at a centre of symmetry and [134134] conformation (196). The PdCl$_2$-complex has two crystallo-graphically independent cations, one of type I [134314] with a molecular mirror plane and the other of type II, with the odd hydrogen on a nitrogen atom and [13444] conformation (197). Although only one structure has been determined for each type of [16]ane-N$_2$S$_2$ macrocycle it is remarkable that they all have different conformations: I [134314], II [13444], III [134134] and V [4444], Figure 3.26.

The 1,4,7,10-[16]ane-N$_4$ macrocycle has been analyzed as the high-spin Ni(II) complex (168). There is a 2-fold axis in the plane of the molecule, so that the type V macrocycle has a [133333] conformation. The coordination produces three 5-membered and a 9-membered chelate ring. The latter has [333] conformation. Two trans chloro groups complete the coordination octahedron, Figure 3.27.

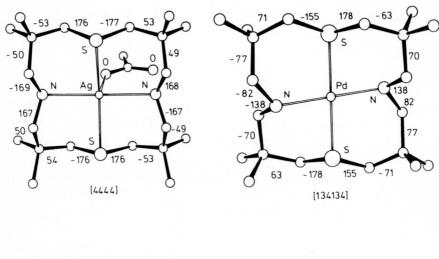

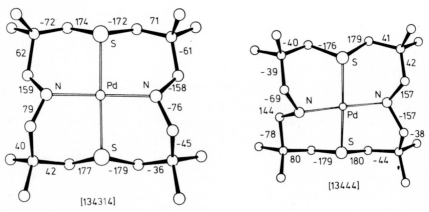

Figure 3.26. Complexes of [16]ane-N_2S_2 with Ag(I) and Pd(II) showing four different types and conformations (196-8). [4444], type V; [134134], type III; [134314], type I; [13444], type II

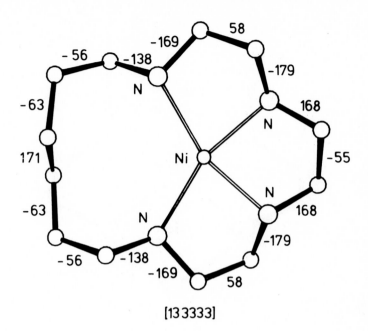

[133333]

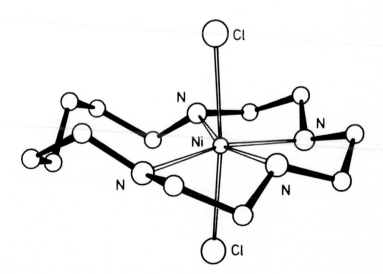

Figure 3.27. Drawings of the [Ni(1,4,7,10-[16]ane-N$_4$)Cl$_2$] structure to illustrate the [133333] conformation and the
coordination geometry (168)

Only one pentadentate [16]ane has been structurally characterized (199). It is correctly formulated as [16]ane-1,4,7,10,13-N_5,

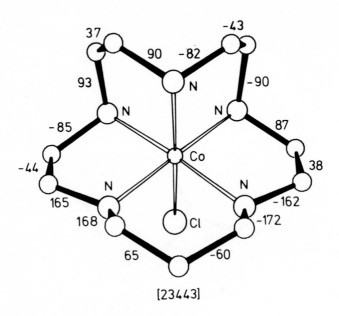

[23443]

In its Co(III) complex it has a [23443] conformation. Nitrogen-7
bends over into an axial position of the octahedral coordination
sphere, completed by an axial chloro donor. This structure shows
remarkable resemblance with an open-chain S_3N_2Co(III) complex
(200), emphasizing the unity of chemistry.

3.4 LARGER MACROCYCLICS

This review is arbitrarily restricted to the stereochemistry of
transition-metal macromonocyclics, for ring sizes not exceeding
sixteen, and excluding crown ethers. In broad perspective it is
however, proper to conclude by reference to some related compounds
beyond the scope of this work, especially to emphasize that the
method of conformational definition is not restricted to the
chosen field of study. The Dale nomenclature, without our
refinement, has for instance been applied to crown ethers (24).
It is pointed out that, in terms of the present system, some ill-
defined non-angular and uniangular forms would not feature. The
method can likewise be extended to the conformational analysis of

cryptands and other macropolycyclic forms, considered as fused macromonocyclics.

Of the macromonocyclics with more than sixteen atoms in the ring, eighteen-membered macrocyclics have been studied in most detail. The symmetrical form of the [18]ane-N_6 ligand (201) has a [4545] conformation while its Co(III) complex of $\bar{3}$ symmetry has [333333] conformation (202). Two structures of [18]ane-N_2S_4 macrocycles have also been reported (203). The unsubstituted macrocycle has a [4545] conformation, whereas the bis(4-chlorobenzoyl)-N substituted ligand has a [133245] conformation (203). Further examples in this class include (Ph)$_4$[18]ane-P_4R_2 macrocycles, where R = O,S,N(C_3H_7) (204).

A few related lanthanide complexes are also known. Two Eu(III) complexes have been reported. In the 1,4,7,10(acetato)-[12]ane-N_4 complex the macrocycle has a square [3333] conformation (94) and is further characterized in Table 2. The [15]ane-O_5 complex of Eu(III) is disordered (205). Neodymium nitrate forms a 1:1 complex with [18]ane-O_6 in the conformation [234243] (206). The samarium complex of [15]ane-O_5 has a macrocyclic conformation of [23424] (207), compared to [2436] for the free macrocycle. The terbium complex of 1,4,8,11-(tetraacetato)cyclam has been structurally investigated (208) and the macrocycle shown to have a [3434] conformation.

4. STERIC EFFECTS OF COORDINATION GEOMETRY

An important aim of stereochemical analysis of macrocyclic metal complexes is to establish the relationship, if any, between coordination geometry and macrocyclic conformation. Since coordination geometry however, is a function of many factors like ring flexibility, macrocyclic configuration, counter-ion effects, cation size and electron configuration of the central ion, meaningful correlations can only be formulated in terms of sufficiently large data sets. Whereas most types are represented by either isolated examples or a small number of structures, detailed comparisons would still be premature. Although not adequate, the data set for fourteen-membered macrocyclics is the only one approaching a meaningful size and the following preliminary correlation is offered with these constraints in mind.

For nine-membered macrocyclics it is noticed that a regular facial coordination goes with a [333] macrocyclic conformation whereas facial coordination to Jahn-Teller distorted metal ions or metal ions which prefer square-planar coordination, correlates with a [234] conformation.

No regular trend relating coordination geometry and the conformation of twelve-membered macrocyclics is evident. There is a weak correlation between cis-octahedral and trigonal bipyramidal geometries with a [2424] conformation and of square pyramidal coordination with [3333]. Other common conformations include [2343] and [2334].

For the purpose of this discussion it is useful to divide the fourteen-membered macrocyclics into 4, 5 and 6-coordinate complexes, in this order.

Two 4-coordinate structures of fourteen-membered macrocyclics with tetrahedral metal ions have been reported. In both cases the macrocycle is sterically prevented from saturating all four coordination sites. It binds instead to two separate metal ions. The macrocycle concerned is [14]ane-S_4 and the metal ions are Cu(I) and Hg(II) respectively (176, 172). In the copper complex there is a 1:1 ratio of metal ion to macrocycle with a conformation of [223223]. Three endo sulfur donors coordinate a single copper ion together with an exo sulfur from another ligand. The mercury:macrocycle ratio is 2:1. Two chloro groups complete the coordination sphere together with two exodentate sulfur donors. For macrocycles with 1,4,8,11-donors square-planar coordination occurs with macrocyclic conformations of [3434], [3344] and [133133]. Square-planar complexes of 1,4,7,11-[14]ane-N_4 have a [2444] conformation.

A general trend is observed for complexes with 5-coordinate metal ions. In square pyramidal complexes where the macrocycle provides a square arrangement of four donors with the metal ion above this coordination plane, but below the apical donor atom, the macrocycle has a [3344] conformation. The D_1-$\hat{M}$-D_2 angles subtended by trans donors in these complexes are close to 180° and differ by less than 5°. For complexes with the metal ion in trigonal bipyramidal environment the macrocycle has a [133133] conformation with the one-bond side in the propylene bridge. In these complexes the two D-$\hat{M}$-D trans angles are very different. They should ideally differ by 60° between 120° and 180°, but larger differences occur. Complexes with intermediate geometries adopt a [3434] macrocyclic conformation. The D-M̂-D angles in these cases differ by about 20°-30°.

For six-coordination the macrocycle either folds to facilitate cis-octahedral coordination or in planar form it allows two further ligands to complete a trans octahedron. In the former class the macrocycle is commonly of type V and of either [124124] or [133133] conformation. The planar macrocycles in trans octahedra adopt conformations of either [3434], [133133] or [3344], the same forms favoured by square-planar coordination.

These arrangements are allowed by macrocycles of types I, III or V. The trans [133133] conformation differs from the cis[133133] in having the one-bond side between the two carbon atoms of the ethylene bridge.

5. CONCLUSION

Macrocyclics represent a distinct class of coordination complexes with metals of all kinds. Transition-metal complexes are the most widely studied and have been emphasized to the exclusion of representative and inner-transition metal complexes in this review. In view of our declared interest in the conformation of macrocycles, these excluded, more ionic complexes, in a sense however, provide a more fertile field of study, since the conformational preference of the central ion is largely eliminated and replaced by a simple principle of closest packing. Review of this topic could be of timely importance, also to explore the relationship between crown ethers and oxygen-free macrocyclics.

The present review re-emphasizes the characteristic flavour imparted to macrocyclic conformation and chemisty by ring-size factors. The average metal ion fits most comfortably into [14]ane cavities and special adjustment is required to form complexes outside the range of twelve- to sixteen-membered macrocyclics. Macrocyclics of smaller ring systems resemble standard coordina-tion more closely in that suitable coordination polyhedra are established by a simple distribution of the polydentate cyclic ligand, together with other available ligands around the central ion. The flexibility of macrocycles, larger than [16]anes likewise enables the most advantageous coordination polyhedra to be established with little or no work against steric factors. The chemically interesting interplay between ionic size factors, conformational rigidity, electronic orbital requirements and steric congestion is therefore manifested only within this relatively narrow special range of compositions. However, most of the macrocyclics of this review are tetradentate and synthesis of more penta- and hexadentate macrocyclics should extend the chemically interesting range towards [22]anes. Future activity in macrocyclic chemistry is anticipated on this front.

A useful scheme, of general validity, for the specification of medium-size macrocyclics has been developed in the course of this review. In principle the method applies directly for larger macrocyclic rings, but as the number of sides increases beyond four, the symbolic description of conformational types becomes less informative. A useful extension of the method to deal with these, more complicated modes of conformational bending, has already been planned. This consists of approximating the macrocyclic rings by straight edges connecting the corners between the conformational sides. A conformation like [133133] can then immediately be described quantitatively in terms of its parameters of puckering, calculated as for six-membered rings. As an illustration of this approach, symbolically equivalent conformational types, obtained for fourteen-membered macrocyclics are compared in terms of the pucker of hexagons defined by their configurational corners, in Table 10. The macrocycle in the trans

TABLE 10
Comparison of four six-edged [14]ane macrocyclics in terms of their conformations simplified to puckered hexagons

Macrocyclic	Conformation	$Q(\text{Å})$	$\theta(°)$	$\phi(°)$	Puckered Hexagon	Coord. Geom.	Ref.
Ru[14]ane-S_4	[124124]$_2$	1.19	90	215	$^2T_4$	cis	174
NiMe$_6$-cyclam	[124124]	1.17	90	252	$^{2,5}B$	cis	134
Ru-cyclam	[133133]2/m	0.58	180	150	$^1C_4$	trans	110
HgMe$_6$-cyclam	[133133]$_2$	1.25	90	356	$^{1,4}B$	cis	136

complex has a clear chair conformation compared to the boat forms of the cis species. The similarity of the Ni and Hg complexes of hexamethylcyclam is also better emphasized by their common boat forms than by their full conformational symbols. This promising procedure is recommended as a standard analysis of large-ring macrocyclic conformations.

During the course of this work several examples of interesting structures that had to be excluded from the review because of our inability to secure suitable crystallographic atomic coordinates, were encountered. This type of wasted effort is readily prevented by either publication of coordinates or suitable archiving, the former of which probably is the more appropriate procedure. This is especially true in macrocyclic chemistry, where molecular conformation should continue to be of increasing importance.

6.   ACKNOWLEDGEMENTS

We sincerely appreciate generous financial grants from the Foundation for Research Development and the University of the Witwatersrand in support of the Structural Chemistry Research Group. The friendly retrieval service of the Cambridge Crystallographic Data Centre is gratefully acknowledged.

7. REFERENCES

1. E. Vogel, M. Köcher, H. Schmickler and J. Lex, Angew. Chem. Int. Ed. Engl., 25 (1986) 257-259.

2. N. F. Curtis, J. Chem. Soc., (1960) 4409-4413.

3. L. Y. Martin, L. J. DeHayes, L. J. Zompa, and D. H. Busch, J. Am. Chem. Soc., 96 (1974) 4046-4048.

4. D. K. Cabbiness and D. W. Margerum, J. Am. Chem. Soc., 92 (1970) 2151-2153.

5. D. H. Busch, K. Farmery, V. Goedken, V. Katovic, A. C. Melnyk, C. R. Sperati and N. Tokel, Adv. Chem. Ser., 100 (1971) 44-78.

6. D. K. Cabbiness and D. W. Margerum, J. Am. Chem. Soc., 91 (1969) 6540-6541.

7. D. C. Olson and J. Vasilevskis, Inorg. Chem., 8 (1969) 1611-1621.

8. G. W. Gokel, D. M. Dishong, R. A. Schultz and V. J. Gatto, Synthesis, 12 (1982) 997-1012.

9. J. J. Christensen, D. J. Eatough and R. M. Izatt, Chem. Rev., 74 (1974) 351-384.

10. R. M. Izatt, J. S. Bradshaw, S. A. Nielsen, J. D. Lamb and J. J. Christensen, Chem. Rev., 85 (1985) 271-339.

11. L. F. Lindoy, Chem. Soc. Rev., 4 (1975) 421-441.

12. R. C. Weast, (Ed), "CRC Handbook of Chemistry and Physics", 5th Edition, CRC Press, (1977) page C-1.

13. G. A. Melson, in "Coordination Chemistry of Macrocyclic Compounds", Ed: G. A. Melson, Plenum Press, New York (1979) p1,17.

14. J. C. A. Boeyens, Stuct. and Bonding, 63 (1985) 65-101.

15. N. F. Curtis, in "Coordination Chemistry of Macrocyclic Compounds", Ed: G. A. Melson, Plenum Press, New York (1979) p219-344.

16. W. Klyne and V. Prelog, Experientia, 16 (1960) 521-523.

17. J. E. Kilpatrick, K. S. Pitzer and R. Spitzer, J. Am. Chem. Soc., 69 (1947) 2483.

18. D. Cremer and J. A. Pople, J. Am. Chem. Soc., 97 (1975) 1354-1358.

19. J. B. Hendrickson, J. Am. Chem. Soc., 89 (1967) 7047-7061.

20. J. C. A. Boeyens, J. Crys. Mol. Struct., 8 (1978) 317-320.

21. D. F. Bocian, H. M. Pickett, T. C. Rounds and H. L. Strauss, J. Am. Chem. Soc., 97 (1975) 687.

22. I. K. Boessenkool and J. C. A. Boeyens, J. Crys. Mol. Struct., 10 (1980) 11-18.

23. J. B. Hendrickson, J. Am. Chem. Soc., 89 (1967) 7036-7043.

24. J. Dale, Acta Chem. Scand., 27 (1973) 1115-1158 and Tetrahedron, 30 (1974) 1683-1694.

25. B. Bosnich, C. K. Poon and M. L. Tobe, Inorg. Chem., 4 (1965) 1102-1108.

26. M. Oki in "Topics in Stereochemistry", Volume 14, Eds: N. L. Allinger, E. L. Eliel and S. H. Wilen, J. Wiley and Sons, New York, p6.

27. V. J. Thöm, C. C. Fox, J. C. A. Boeyens and R. D. Hancock, J. Am. Chem. Soc., 106 (1984) 5947-5955.

28. J. C. A. Boeyens, Acta Cryst., C39 (1983) 846-849.

29. J. C. A. Boeyens, R. D. Hancock and V. J. Thöm, J. Cryst. Spectr. Res., 14 (1984) 261-268.

30. W. Clegg, P. Leupin, D. T. Richens, A. G. Sykes and E. S. Raper, Acta Cryst., C41 (1985) 530-532.

31. L. Fabbrizzi, C. Mealli and P. Paoletti, J. Chem. Res., (1979) 170-171.

32. V. B. Pett, L. L. Diaddario, E. R. Dockal, P. W. Corfield, C. Ceccarelli, M. D. Glick, L. A. Ochrymowycz and D. B. Rorabacher, Inorg. Chem., 22 (1983) 3661-3670.

33. M. Kato and T. Ito, Inorg. Chem., 24 (1985) 509-514.

34. T. C. W. Mak, C.-M. Che and K.-Y. Wong, J. Chem. Soc., Chem. Comm. (1985) 986-988.

35. R. S. Glass, G. S. Wilson and W. N. Setzer, J. Am. Chem. Soc., 102 (1980) 5068-5069.

36. W. N. Setzer, C. A. Ogle, G. S. Wilson and R. S. Glass, Inorg. Chem., 22 (1983) 266-271.

37. K. Wieghardt, H.-J. Küppers and J. Weiss, Inorg. Chem., 24 (1985) 3067-3071.

38. H.-J. Küppers, A. Neves, C. Pomp, D. Ventur, K. Wieghardt, B. Nuber and J. Weiss, Inorg. Chem., 25 (1986) 2400-2408.

39. M. T. Ashby and D. L. Lichtenberger, Inorg. Chem., 24 (1985) 636-638.

40. S. M. Hart, J. C. A. Boeyens, J. P. Michael and R. D. Hancock, J. Chem. Soc., Dalton Trans. (1983) 1601-1606.

41. J. C. A. Boeyens, S. M. Dobson and R. D. Hancock, Inorg. Chem., 24 (1985) 3073-3076.

42. S.M. Dobson, Ph.D. Thesis (1986) University of the Witwatersrand, Johannesburg.

43. L. J. Zompa and T. N. Margulis, Inorg. Chim. Acta, 28 (1978) L157-L159.

44. R. D. Bereman, M. R. Churchill, P. M. Schaber and M. E. Winkler, Inorg. Chem., 18 (1979) 3122-3125.

45. W. F. Schwindinger, T. G. Fawcett, R. A. Lalancette, J. A. Potenza and H. J. Schugar, Inorg. Chem., 19 (1980) 1379-1381.

46. K. Wieghardt, M. Köppen, W. Swiridoff and J. Weiss, J. Chem. Soc., Dalton Trans. (1969) 1869-1872.

47. S. Sato, S. Ohba, S. Shimba, S. Fujinami, M. Shibata and Y. Saito, Acta Cryst., B36 (1980) 43-47.

48. J. C. A. Boeyens, A. G. S. Forbes, R. D. Hancock and K. Wieghardt, Inorg. Chem., 24 (1985) 2926-2931.

49. K. Wieghardt, W. Walz, B. Nuber, J. Weiss, A. Ozarowski, H. Stratemeier and D. Reinen, Inorg. Chem., 25 (1986) 1650-1654.

50. P. Chaudhuri, K. Wieghardt, Y.-H. Tsai and C. Krüger, Inorg. Chem., 23 (1984) 427-432.

51. K. Wieghardt, W. Schmidt, B. Nuber and J. Weiss, Chem. Ber, 112 (1979) 2220-2230.

52. K. Wieghardt, W. Schmidt, B. Nuber, B. Prikner and J. Weiss, Chem. Ber, 113 (1980) 36-41.

53. K. Wieghardt, W. Schmidt, R. van Eldik, B. Nuber and J. Weiss, Inorg. Chem., 19 (1980) 2922-2926.

54. K. Wieghardt, K. Pohl and W. Gebert, Angew. Chem. Int. Ed. Engl., 22 (1983) 727.

55. K. Wieghardt, M. Hahn, W. Swiridoff and J. Weiss, Angew. Chem. Int. Ed. Engl., 22 (1983) 491-492.

56. K. Wieghardt, M. Hahn, W. Swiridoff and J. Weiss, Inorg. Chem., 23 (1984) 94-99.

57. K. Wieghardt, U. Bossek, K. Volckmar, W. Swiridoff and J. Weiss, Inorg. Chem., 23 (1984) 1387-1389.

58. K. Wieghardt, W. Herrmann, M. Köppen, I. Jibril and G.

Huttner, Z. Naturforsch., 39b (1984) 1335-1343.

59. P. Chaudhuri, K. Wieghardt, I. Jibril and G. Huttner, Z. Naturforsch., 39b (1984) 1172-1176.

60. K. Wieghardt, U. Bossek, D. Ventur and J. Weiss, J. Chem. Soc., Chem. Comm. (1985) 347-349.

61. K. Wieghardt, W. Schmidt, H. Endres and C. R. Wolfe, Chem. Ber, 112 (1979) 2837-2846.

62. K. Wieghardt, U. Bossek and W. Gebert, Angew. Chem. Int. Ed. Engl., 22 (1983) 328-329.

63. K. Wieghardt, K. Pohl, I. Jibril and G. Huttner, Angew. Chem. Int. Ed. Engl., 23 (1984) 77-78.

64. K. Wieghardt, C. Pomp, B. Nuber and J. Weiss, Inorg. Chem., 25 (1986) 1659-1661.

65. K. Wieghardt, M. Kleine-Boymann, B. Nuber and J. Weiss, Inorg. Chem., 25 (1986) 1654-1659.

66. K. Wieghardt, M. Kleine-Boymann, B. Nuber, J. Weiss, L. Zsolnai and G. Huttner, Inorg. Chem., 25 (1986) 1647-1650.

67. M. Mikami, R. Kuroda, M. Konno and Y. Saito, Acta Cryst., B33 (1977) 1485-1489.

68. K. Wieghardt, P. Chaudhuri, B. Nuber and J. Weiss, Inorg. Chem., 21 (1982) 3086-3090.

69. K. Wieghardt, G. Backes-Dahmann, W. Hermann and J. Weiss, Angew. Chem. Int. Ed. Engl., 23 (1984) 899-900.

70. P. Chaudhuri, K. Wieghardt, B. Nuber and J. Weiss, J. Chem. Soc., Chem. Comm. (1985) 265-266.

71. P. Chaudhuri, M. Guttner, D. Ventur, K. Wieghardt, B. Nuber and J. Weiss, J. Chem. Soc., Chem. Comm. (1985) 1618-1620.

72. K. Wieghardt, M. Kleine-Boymann, B. Nuber and J. Weiss, Inorg. Chem., 25 (1986) 1309-1313.

73. K. Wieghardt, U. Bossek, P. Chaudhuri, W. Herrmann, B. C. Menke and J. Weiss, Inorg. Chem., 21 (1982) 4308-4313.

74. M. J. van der Merwe, J. C. A. Boeyens and R. D. Hancock, Inorg. Chem., 22 (1983) 3489-3490.

75. K. Wieghardt, U. Bossek, M. Guttman and J. Weiss, Z. Naturforsch., 38b (1983) 81-89.

76. S. G. Taylor, M. R. Snow and T. W. Hambley, Aust. J. Chem., 36 (1983) 2359-2368.

77. L. J. Zompa and T. N. Margulis, Inorg. Chim. Acta, 45 (1980) L263-L264.

78. H. Gampp, M. M. Roberts and S. J. Lippard, Inorg. Chem., 23 (1984) 2793-2798.

79. B. N. Diel, R. C. Haltiwanger and A. D. Norman, J. Am. Chem. Soc., 104 (1982) 4700-4701.

80. P. G. Jones, T. Gries, H. Grützmacher, H. W. Roesky, J. Schimkowiak and G. M. Sheldrick, Angew. Chem. Int. Ed. Engl., 23 (1984) 376-376.

81. R. Clay, P. Murray-Rust and J. Murray-Rust, Acta Cryst., B35 (1979) 1894-1895.

82. Y. Iitaka, M. Shina and E. Kimura, Inorg. Chem., 13 (1974) 2886-2891.

83. N. Matsumoto, A. Hirano, T. Hara and A. Ohyoshi, J. Chem. Soc., Dalton Trans. (1983) 2405-2410.

84. J. H. Loehlin and E. B. Fleischer, Acta Cryst., B32 (1976) 3063-3066.

85. M. Ciampolini, M. Micheloni, N. Nardi, P. Paoletti, P. Dapporto and F. Zanobini, J. Chem. Soc., Dalton Trans. (1984) 1357-1362.

86. J. Giusti, S. Chimichi, M. Ciampolini, M. Sabat and D. Masi, Inorg. Chim. Acta, 88 (1984) 51-54.

87. J. H. Coates, D. M. M. A. Hadi, T. W. Hambley, S. F. Lincoln and J. R. Rodgers, Cryst. Struct. Comm., 11 (1982) 815.

88. T. Sakurai, S. Tsuboyama and K. Tsuboyama, Acta Cryst., B36 (1980) 1797-1801.

89. T. Sakurai, K. Kobayashi, A. Hasegawa, S. Tsuboyama and K. Tsuboyama, Acta Cryst., B38 (1982) 107-111.

90. T. Sakurai, K. Kobayashi, K. Tsuboyama and S. Tsuboyama, Acta Cryst., B34 (1978) 1144-1148.

91. S. Buøen, J. Dale, P. Groth and J. Krane, J. Chem. Soc., Chem. Comm. (1982) 1172-1174.

92. P. Groth, Acta Chem. Scand., A38 (1984) 342-344.

93. A. Riesen, M. Zehnder and T. A. Kaden, J. Chem. Soc., Chem. Comm. (1985) 1336-1338.

94. M.-R. Spirlet, J. Rebizant, J. F. Desreux and M.-F. Loncin, Inorg. Chem., 23 (1984) 359-363.

95. T. Sakurai, K. Kobayashi, K. Tsuboyama and S. Tsuboyama, Acta Cryst., B34 (1978) 3465-3469.

96. T. Sakurai, Y. Watanabe, K. Tsuboyama and S. Tsuboyama,

Acta Cryst., B37 (1981) 613-618.

97. H. Hiramatsu, T. Sakurai, K. Tsuboyama and S. Tsuboyama, Acta Cryst., B35 (1979) 1241-1244.

98. T. Sakurai, H. Hiramatsu, K. Tsuboyama and S. Tsuboyama, Acta Cryst., B36 (1980) 2453-2456.

99. S. Tsuboyama, K. Kobayashi, T. Sakurai and K. Tsuboyama, Acta Cryst., C40 (1984) 1178-1181.

100. T. Sakurai, K. Kobayashi, H. Masuda, S. Tsuboyama and K. Tsuboyama, Acta Cryst., C39 (1983) 334-337.

101. K. Kobayashi, T. Sakurai, A. Hasegawa, S. Tsuboyama and K. Tsuboyama, Acta Cryst., B38 (1982) 1154-1158.

102. J. M. Waters and K. R. Whittle, J. Inorg. Nucl. Chem., 34 (1972) 155-161.

103. C. Nave and M. R. Truter, J. Chem. Soc., Dalton Trans. (1974) 2351-3454.

104. P. A. Tasker and L. Sklar, J. Crys. Mol. Struct., 5 (1975) 329.

105. B. Bosnich, R. Mason, P. J. Pauling, G. B. Robertson and M. L. Tobe, J. Chem. Soc., Chem. Comm. (1965) 97-98.

106. N. W. Alcock, N. Herron and P. Moore, J. Chem. Soc., Dalton Trans. (1979) 1486-1491.

107. T. F. Lai and C. K. Poon, Inorg. Chem., 15 (1976) 1562-1566.

108. C.-M. Che, S.-S. Kwong, C. K. Chung and T.-F. Lai, Inorg. Chem., 24 (1985) 1359-1363.

109. E. Forsellini, T. Parasassi, G. Bombieri, M. L. Tobe and M. E. Sosa, Acta Cryst., C42 (1986) 563-565.

110. D. D. Walker and H. Taube, Inorg. Chem., 20 (1981) 2828-2834.

111. S. A. Zuckman, G. M. Freeman, D. E. Troutner, W. A. Volkert, R. A. Holmes, D. G. Van Derveer and E. K. Barefield, Inorg. Chem., 20 (1981) 2386-2389.

112. T. Ito, H. Ito and K. Toriumi, Chem. Lett., (1981) 1101-1104.

113. T. Ito, M. Sugimoto, K. Toriumi and H. Ito, Chem. Lett., (1981) 1477-1478.

114. E. Bang and O. Mønsted, Acta Chem. Scand., A36 (1982) 353-359.

115. A. W. Addison and E. Sinn, Inorg. Chem., 22 (1983)

98

1225-1228.

116. M. Kato and T. Ito, Bull. Chem. Soc. Jpn., 59 (1986)
285-294.

117. M. Yamashita, H. Ito, K. Toriumi and T. Ito, Inorg. Chem.,
22 (1983) 1566-1568.

118. L. Prasad and A. McAuley, Acta Cryst., C39 (1983)
1175-1177.

119. J. F. Endicott, J. Lilie, J. M. Kuszaj, B. S. Ramaswamy, W.
G. Schmonsees, M. G. Simic, M. D. Glick and D. P. Rillema,
J. Am. Chem. Soc., 99 (1977) 429-439.

120. K. P. Larsen and H. Toftlund, Acta Chem. Scand., A31 (1977)
182-186.

121. Z. Urbanczyk-Lipkowska, J. W. Krajewski, P. Gluzinski, G.
D. Andreetti and G. Bocelli, Acta Cryst., B37 (1981)
470-473.

122. J. Krajewski, Z. Urbanczyk-Lipkowska and P. Gluzinski,
Bull. Acad. Pol. Sci. Sci. Chem., 25 (1977) 853.

123. R. J. Restivo, G. Ferguson, R. W. Hay and D. P. Piplani, J.
Chem. Soc., Dalton Trans. (1978) 1131-1134.

124. J. W. Krajewski, Z. Urbanczyk-Lipkowska and P. Gluzinski,
Pol. J. Chem, 52 (1978) 1513.

125. R. W. Hay, B. Jeragh, G. Ferguson, B. Kaitner and B. L.
Ruhl, J. Chem. Soc., Dalton Trans. (1982) 1531-1539.

126. J. W. Krajewski, Z. Urbanczyk-Lipkowska and P. Gluzinski,
Pol. J. Chem., 54 (1980) 2189.

127. T.-J. Lee, T.-Y. Lee, W.-B. Juang and C.-S. Chung, Acta
Cryst., C41 (1985) 1596-1598.

128. T.-J. Lee, T.-Y. Lee, W.-B. Juang and C.-S. Chung, Acta
Cryst., C41 (1985) 1745-1748.

129. T. Ito, H. Ito and K. Toriumi, Acta Cryst., B37 (1981)
1412-1415.

130. T. H. Lu, T. J. Lee and C. S. Chung, Acta Cryst., A37
(1981) C231.

131. J. W. Krajewski, Z. Urbanczyk-Lipkowska and P. Gluzinski,
Cryst. Struct. Commun., 6 (1977) 817.

132. B. H. Toby, J. L. Hughey, T. G. Fawcett, J. A. Potenza and
H. J. Schugar, Acta Cryst., B37 (1981) 1737-1739.

133. E. Zeigerson, I. Bar, J. Bernstein, L. J. Kirschenbaum and
D. Meyerstein, Inorg. Chem., 21 (1982) 73-80.

134. H. Ito, J. Fujita, K. Toriumi and T. Ito, Bull. Chem. Soc. Jpn., 54 (1981) 2988-2994.

135. H. Ito, M. Sugimoto and T. Ito, Bull. Chem. Soc. Jpn., 55 (1982) 1971-1972.

136. M. R. Burke and M. F. Richardson, Inorg. Chim. Acta, 69 (1983) 29-35.

137. T.-J. Lee, H. Y. J. Lee, C.-S. Lee and C. S. Chung, Acta Cryst., C40 (1984) 641-644.

138. K. Toriumi and T. Ito, Acta Cryst., B37 (1981) 240-243.

139. N. F. Curtis, D. A. Swann and T. N. Waters, J. Chem. Soc., Dalton Trans. (1973) 1408-1413.

140. J. L. Hughey, T. G. Fawcett, S. M. Rudich, R. A. Lalancette, J. A. Potenza and H. J. Schugar, J. Am. Chem. Soc., 101 (1979) 2617-2623.

141. P. Gluzinski, J. W. Krajewski and Z. Urbanczyk-Lipkowska, Acta Cryst., B36 (1980) 1695-1698.

142. E.-I. Ochiai, S. Rettig and J. Trotter, Can. J. Chem., 56 (1978) 267-272.

143. T. Ito and K. Toriumi, Acta Cryst., B37 (1981) 88-92.

144. T. Ito, K. Toriumi and H. Ito, Bull. Chem. Soc. Jpn., 54 (1981) 1096-1100.

145. R. Clay, J. Murray-Rust and P. Murray-Rust, J. Chem. Soc., Dalton Trans. (1979) 1135-1139.

146. N. F. Curtis, D. A. Swann and T. N. Waters, J. Chem. Soc., Dalton Trans. (1973) 1963-1974.

147. P. O. Whimp, M. F. Bailey and N. F. Curtis, J. Chem. Soc., A (1970) 1956-1963.

148. K. B. Mertes, Inorg. Chem., 17 (1978) 49-52.

149. R. Temple, D. A. House and W. T. Robinson, Acta Cryst., C40 (1984) 1789-1791.

150. E. Bang and O. Mønsted, Acta Chem. Scand., A38 (1984) 281-287.

151. R. A. Bauer, W. R. Robinson and D. W. Margerum, J. Chem. Soc., Chem. Comm. (1973) 289-290.

152. R. M. Smith and A. E. Martell, "Critical Stability Constants.", Volume 2: Amines (1975) and Volume 5: First Supplement (1982) Plenum Press; New York.

153. F. Wagner, M. T. Mocella, M. J. D'Aniello, A. H.-J. Wang and E. K. Barefield, J. Am. Chem. Soc., 96 (1974)

100

2625-2627.

154. N. W. Alcock, N. Herron and P. Moore, J. Chem. Soc., Dalton Trans. (1978) 1282-1288.

155. M. J. D'Aniello, M. T. Mocella, F. Wagner, E. K. Barefield and I. C. Paul, J. Am. Chem. Soc., 97 (1975) 192-194.

156. K. D. Hodges, R. G. Wollman, S. L. Kessel, D. N. Hendrickson, D. G. Van Deveer and E. K. Barefield, J. Am. Chem. Soc., 101 (1979) 906-917.

157. T. W. Hambley, J. Chem. Soc., Dalton Trans. (1986) 565-569.

158. I. S. Crick, R. W. Gable, B. F. Hoskins and P. A. Tregloan, Inorg. Chim. Acta, 111 (1986) 35-38.

159. C.-M. Che, K.-Y. Wong and T. C. W. Mak, J. Chem. Soc., Chem. Comm. (1985) 988-990.

160. E. K. Barefield, G. M. Freeman and D. G. Van Deveer, J. Chem. Soc., Chem. Comm. (1983) 1358-1360.

161. G. M. Freeman, E. K. Barefield and D. G. Van Deveer, Inorg. Chem., 23 (1984) 3092-3103.

162. I. Murase, M. Mikuriya, H. Sonoda and S. Kida, J. Chem. Soc., Chem. Comm. (1984) 692-694.

163. I. Murase, M. Mikuriya, H. Sonoda, Y. Fukuda and S. Kida, J. Chem. Soc., Dalton Trans. (1986) 953-959.

164. X. Ji-De, N. Shi-Sheng and L. Yu-Juan, Inorg. Chim. Acta, 111 (1986) 61-65.

165. E. K. Barefield, D. Chueng, D. G. Van Deveer and F. Wagner, J. Chem. Soc., Chem. Comm. (1981) 302-304.

166. R. Schneider, A. Riesen and T. A. Kaden, Helv. Chim. Acta, 68 (1985) 53-61.

167. N. W. Alcock, R. G. Kingston, P. Moore and C. Pierpoint, J. Chem. Soc., Dalton Trans. (1984) 1937-1943.

168. M. Sugimoto, J. Fujita, H. Ito, K. Toriumi and T. Ito, Inorg. Chem., 22 (1983) 955-960.

169. M. Sugimoto, H. Ito, K. Toriumi and T. Ito, Acta Cryst., B38 (1982) 2453-2455.

170. A. Bianchi, L. Bologni, P. Dapporto, M. Micheloni and P. Paoletti, Inorg. Chem., 23 (1984) 1201-1205.

171. R. E. DeSimone and M. D. Glick, J. Am. Chem. Soc., 98 (1976) 762-767.

172. N. W. Alcock, N. Herron and P. Moore, J. Chem. Soc., Dalton Trans. (1978) 394-399.

173. N. Galesic, M. Herceg and D. Sevdic, Acta Cryst., C42 (1986) 565-568.

174. T.-F. Lai and C.-K. Poon, J. Chem. Soc., Dalton Trans. (1982) 1465-1469.

175. M. D. Glick, D. P. Gavel, L. L. Diaddario and D. B. Rorabacher, Inorg. Chem., 15 (1976) 1190-1193.

176. E. R. Dockal, L. L. Diaddario, M. D. Glick and D. B. Rorabacher, J. Am. Chem. Soc., 99 (1977) 4530-4532.

177. P. H. Davis, L. K. White and R. L. Belford, Inorg. Chem., 14 (1975) 1753-1757.

178. R. E. DeSimone and M. D. Glick, J. Coord. Chem., 5 (1976) 181.

179. T. Yoshida, T. Ueda, T. Adachi, K. Yamamoto and T. Higuchi, J. Chem. Soc., Chem. Comm. (1985) 1137-1138.

180. N. K. Dalley, J. S. Smith, S. B. Larson, K. L. Matheson, J. J. Christensen and R. M. Izatt, J. Chem. Soc., Chem. Comm. (1975) 84 and N. K. Dalley, J. S. Smith, S. B. Larson, J. J. Christensen and R. M. Izatt, J. Chem. Soc., Chem. Comm. (1975) 43.

181. R. Bartsch, S. Hietkamp, S. Morton, H. Peters and O. Stelzer, Inorg. Chem., 22 (1983) 3624-3632.

182. H. Hope, M. Viggiano, B. Moezzi and P. P. Power, Inorg. Chem., 23 (1984) 2550-2552.

183. N. K. Dalley, S. B. Larson, J. S. Smith, K. L. Matheson, R. M. Izatt and J. J. Christensen, J. Heterocycl. Chem, 18 (1981) 463.

184. P. R. Louis, Y. Agnus and R. Weiss, Acta Cryst., B33 (1977) 1418-1421.

185. P. R. Louis, D. Pelissard and R. Weiss, Acta Cryst., B32 (1976) 1480-1485.

186. P. R. Louis, B. Metz and R. Weiss, Acta Cryst., B30 (1974) 774-780.

187. P. R. Louis, Y. Agnus and R. Weiss, Acta Cryst., B35 (1979) 2905-2910.

188. P. R. Louis, D. Pelissard and R. Weiss, Acta Cryst., B30 (1974) 1889-1894.

189. W. L. Smith, J. D. Ekstrand and K. N. Raymond, J. Am. Chem. Soc., 100 (1978) 3539-3544.

190. N. W. Alcock, E. H. Curzon and P. Moore, J. Chem. Soc.,

Dalton Trans. (1984) 2813-2820.

191. N. W. Alcock, E. H. Curzon, P. Moore and C. Pierpoint, J. Chem. Soc., Dalton Trans. (1984) 605-610.

192. T. E. Jones, L. S. W. L. Sokol, D. B. Rorabacher and M. D. Glick, J. Chem. Soc., Chem. Comm. (1979) 140-141.

193. J. Jr. Cragel, V. B. Pett, M. D. Glick and R. E. DeSimone, Inorg. Chem., 17 (1978) 2885-2893.

194. R. E. DeSimone, J. Cragel Jr, W. H. Ilsley and M. D. Glick, J. Coord. Chem., 9 (1979) 167.

195. R. E. DeSimone and M. D. Glick, Inorg. Chem., 17 (1978) 3574-3577.

196. R. McCrindle, G. Ferguson, A. J. McAlees, M. Parvez and D. K. Stephenson, J. Chem. Soc., Dalton Trans. (1982) 1291-1296.

197. G. Ferguson, R. McCrindle, A. J. McAlees, M. Parvez and D. K. Stephenson, J. Chem. Soc., Dalton Trans. (1983) 1865-1868.

198. G. Ferguson, R. McCrindle and M. Parvez, Acta Cryst., C40 (1984) 354-356.

199. G. Bombieri, E. Forsellini, A. del Pra, C. J. Cooksey, M. Humanes and M. L. Tobe, Inorg. Chim. Acta, 61 (1982) 43-49.

200. J. D. Korp, I. Bernal and J. H. Worrell, Polyhedron, 2 (1983) 323-330.

201. T. N. Margulis and L. J. Zompa, Acta Cryst., B37 (1981) 1426-1428.

202. Y. Yoshikawa, K. Toriumi, T. Ito and H. Yamatera, Bull. Chem. Soc. Jpn., 55 (1982) 1422-1424.

203. H. L. Ammon, K. Chandrasekhar, S. K. Bhattacharjee, S. Shinkai and Y. Honda, Acta Cryst., C40 (1984) 2061-2064.

204. M. Ciampolini, N. Nardi, F. Zanobini, R. Cini and P. L. Orioli, Inorg. Chim. Acta, 76 (1983) L17-L19.

205. J.-C. G. Bünzli, B. Klein, G. Chapuis and K. J. Schenk, Inorg. Chem., 21 (1982) 808-812.

206. J.-C. G. Bünzli, B. Klein and D. Wessner, Inorg. Chim. Acta, 44 (1980) L147-L149.

207. T. J. Lee, H.-R. Sheu, T. I. Chiu and C. T. Chang, Acta Cryst., C39 (1983) 1357-1360.

208. M.-R. Spirlet, J. Rebizant, M.-F. Loncin and J. F. Desreux, Inorg. Chem., 23 (1984) 4278-4283.

Chapter 2

THERMODYNAMIC AND STEREOCHEMICAL ASPECTS OF THE MACROCYCLIC AND
CRYPTATE EFFECTS

H.-J. Buschmann

1   INTRODUCTION

With the discovery of crown ethers and their ability to form
stable complexes with alkali and alkaline-earth cations in solution
(ref. 1) a new chapter in co-ordination chemistry opened; the
previous chapter dealt with chelate complexes, research for which
was started in 1893 by Werner (ref. 2). In 1952 Schwarzenbach first
used the term 'chelate effect' (ref. 3) to provide a justification
for the higher stability constants of the chelate complexes in comparison with their monomolecular analogues. Much experimental work
was done to find the thermodynamic origin of this effect (ref. 4).

Only one year after Werner discovered chelate compounds the
first crown ether was synthesized by Vorländer. He reported the
synthesis of a compound which nowadays would be given the name
"macrocyclic polyether tetraester" (ref. 5). Nearly 40 years later
(1935) another syntheses of crown compounds was reported by
Carothers (ref. 6). In the same year he published an article with
the title "Macrocyclic Esters" (ref. 7). However, the possibility
of these compounds forming stable complexes with alkali and other
ions was not detected then.

The polymerisation of ethyleneoxide in the presence of catalysts
led to the discovery of an unknown cyclic tetramer 1955 (ref. 8).
This substance is now well known as 12-crown-4.

More examples of the preparation of similar polyether compounds
are readily available in the relevant literature (ref. 9). Even the
dissolution of alkali metals in different ethers (ref. 10) did not
stimulate the investigation of this phenomenon, though a complex
formation between metal and solvent molecules seems to be obvious.
This situation changed as a result of the discovery of the natural
occurance of antibiotics which were able to transport ions through
membranes (ref. 11). At this time synthetic analogues of these
antibiotics aroused particular interest.

Only by luck did Pedersen synthesize such compounds and realize
they had the ability to complex alkali ions. He and Carothers had

been working for the same company, mutually unaware of each other's experiments.

To avoid the very cumbersome names obtained by the usage of the IUPAC rules for the macrocyclic polyethers, Pedersen suggested a special "crown" nomenclature (ref. 1). Normally, these names are now used.

The solid crown ether complexes commonly were of 1:1 stoichiometry but different compositions were also found. These crown ethers attracted more and more interest as a result of their ability to act as neutral ligands to complex cations. Which resulted in investigations of their reactions in solution with different cations using a calorimetric titration technique (ref. 12), potentiometric measurements with ion-selective electrodes (ref. 13) and spectroscopic methods (ref. 14).

The development of a further class of neutral ligands started with the synthesis of macrocyclic diamines in the year 1968 (ref. 15). One year later Lehn reported the synthesis of macrobicyclic diamines with oxygen donor atoms containing bridges between both nitrogen atoms (ref. 16). Because it is possible for these ligands to encapsulate cations the name cryptand was suggested. The complexes of cryptands are called cryptates.

The complexes formed with cryptands are generally several orders of magnitude more stable than those of crown ether complexes. The results of the first quantitative measurements of the stability of cryptates was reported shortly after the announcement of their synthesis (ref. 17).

Since that time, the number of publications dealing with crown ethers and cryptands has increased from year to year. In 1982 Gokel and Korzeniowski published a book about the synthesis of macrocyclic polyethers (ref. 18). Their compilation of mono and bicyclic ligands already listed more than 2100 ligands. In the meantime this number will surely have increased even further.

In 1985 Izatt and his co-workers tabulated the known thermodynamic data for cation-macrocyclic interactions (ref. 19). At that time the complexation properties of almost 260 different ligands were given. Thus, the properties of 90 % of the known ligands have not yet been estimated.

1.1 THE MACROCYCLIC AND CRYPTATE EFFECT

In analogy to the 'chelate effect' a 'macrocyclic effect' was first reported by Margerum (ref. 20) for the complexation reactions of macrocyclic tetramines with Cu^{2+}. He compared the measured stability constants with those for noncyclic tetramines. An enthalpic origin of the macrocyclic effect was deduced from calorimetric studies by the same author (ref. 21). However, Paoletti and his coworkers later reported the results of copper-polyamine complexes (ref. 22) and together with another group (ref. 23) they found a dominant contribution of a favourable entropic term to the macrocyclic effect. Paoletti's group continued to study the reactions of different linear and macrocyclic polyamine ligands and their results have been summarized in a short review (ref. 24).

On comparing the results for the complexation of Cu^{2+} by open-chain and cyclic polythiaethers an assignment of the macrocyclic effect to entropic factors was reported (ref. 25). This effect was also observed in the case of polyether ligands. When studying the complex formation of 18-crown-6 and pentaglyme, Frensdorff hypothesized that noncyclic ligands are unable to surround the cation completely because of electrostatic repulsion on the part of the terminal oxygen atoms and an unfavourable entropy contribution occurs when the ligand wraps around the cation (ref. 13). Kodama and Kimura attributed the origin of the macrocyclic effect to favourable entropic contributions (ref. 26). They studied the reaction of 18-crown-6 and tetraglyme with Pb^{2+}; however, the number of donor atoms of both ligands are not identical.

A different result was obtained from the reactions of Na^+, K^+ and Ba^{2+} with pentaethylene glycol, pentaglyme and 18-crown-6 (ref. 27). From the measured data, Izatt and his co-workers concluded that the macrocyclic effect in the systems studied was primarily the result of favourable enthalpic factors.

The interpretation of experimental results became more inconsistent from the moment when reactions with ligands with different donor atoms were investigated. Frensdorff reported long ago that the stability constants for the complexation of Ag^+ by a diaza crown ether and its noncyclic analogues were of comparable magnitude (ref. 13).

The reaction of Cu^{2+} with a macrocyclic ligand containing O, N and S donor atoms was favoured by enthalpic and entropic contributions when compared with an identical noncyclic ligand (ref. 28). No macrocyclic effect was observed in the reaction of Pb^{2+} with the same ligands. The values of the reaction enthalpies and entropies for both ligands were almost identical. On comparison of the stability constants of several linear oxygen and sulphur content ligands with their macrocyclic analogues, towards Ag^+ and Hg^{2+}, no macrocyclic effect was observed.

Neither a macrocyclic nor a significant cryptate effect was observed for the reaction of Ag^+, Hg^{2+} and Cd^{2+} with 1,8-diamino-3,6-dioxaoctane, diaza-18-crown-6 and the cryptand (222) (ref. 30). In the case of the complexation of Sr^{2+} and Ba^{2+} by these monocyclic and bicyclic ligands an enthalpic origin of the cryptate effect was deduced from the data measured. In accordance with this interpretation, Lehn et al also confirmed the dominant contribution of the reaction enthalpy to the cryptate effect (ref. 31); however, they compared the results of the reaction of K^+ with dicyclohexyl-18-crown-6 and the cryptand (222). Their treatment may be doubtful because both ligands possed different kinds of donor atoms.

Only a short review of the older important experimental results is presented. For more detailed information and discussion of these and subsequent data, the corresponding parts of other existing review articles are recommended (ref. 32).

Up to now there is little understanding of many questions about the origin of the macrocyclic and cryptate effect. This seems surprising because, as mentioned before, the complexation properties of many different ligands have already been investigated. To obtain information about the macrocyclic and cryptate effect it is necessary to know the stability constant of the complex formed and the corresponding reaction enthalpy. Further analysis of the review of Izatt et al dealing with the known thermodynamic data for cation-macrocyclic interactions leads to results that are somewhat surprising (ref. 19). The complexation properties of nearly 260 ligands have been studied; thus, 3000 values of stability constants for the reactions of these ligands with different cations in several solvents and mixtures of solvents are known. However, less than 600 values for the reaction enthalpies are summarized. This demonstrates the relatively small basis of data published up to early 1984, used in the discussion about the macrocyclic and cryptate effect.

In this text is not relevant to cite every article dealing with the complexation of any crown ether or cryptand with a charged ion.

2   CATION-LIGAND INTERACTIONS

The interactions between alkali and alkaline-earth cations and ligand donor atoms are mainly of electrostatic nature. In contrast to these 'hard' ions bonds with varying covalent character are formed with 'soft' cations.

Pure electrostatic models have been used for the treatment of cation hydration (ref. 33) and complexation (ref. 34). Such a model was also suggested for the interactions of cations with macrocyclic ligands (ref. 35). In the meantime methods of quantum chemistry have already been used to calculate the bonding of different cations by macrocyclic ethers (ref. 36, 37, 38) and ionophores (ref. 39) that occur naturally.

However, a pure electrostatic model has some advantages in the discussion of complexation reactions. It allows a separate evaluation of various effects influencing these reactions. The sum of the contributions listed should mainly account for the complex stabilities, K, observed as a result of the following fundamental equations:

$$\Delta G = - RT \ln K \qquad (1)$$
and
$$\Delta G = \Delta H - T\Delta S \qquad (2)$$

I. Contributions to the complexation enthalpy
a) part or complete desolvation of the cation
b) change of the ligand solvation
c) interactions between the cation and some or all donor atoms of the ligand
d) repulsion between neighbouring donor atoms
e) influence of the complex formation on the cation interactions with solvent molecules outside the first solvation shell
f) sterical deformations of the ligand due to the complexed cation

II. Contributions to complexation entropy
a) solvation entropy of the cation
b) solvation entropy of the ligand
c) changes in the ligand internal entropy due to orientation, rigidification and conformational changes
d) variation of the number of particles during the reaction
e) changes in translational entropy

It is not intended that the factors mentioned should be discussed in great detail. For further information references to relevant literature will be given.

The solvation and associated desolvation process of a cation
(Ia + IIa) can be described by the Born equation (ref. 40). The
molar ionic solvation Gibbs free energy ΔG_m is given by the following equation

$$\Delta G_m = - \frac{e^2 z^2 A}{2a} \left(1 - \frac{1}{\varepsilon}\right) \tag{3}$$

the charge of a proton being e, the formal charge of the ion Z, the
Avogadro constant A, the relative permittivity of the solvent ε and
the radius of the complexed cation a. The separation of the Born
equation into one enthalpic and one entropic term is possible:

$$\Delta H = - \frac{e^2 z^2 A}{2a} \left[1 - \frac{1}{\varepsilon} - \frac{T}{\varepsilon^2} \left(\frac{\partial \varepsilon}{\partial T}\right)_p \right] \tag{4}$$

and

$$\Delta S = \frac{e^2 z^2 A}{2a} \frac{1}{2} \left(\frac{\partial \varepsilon}{\partial T}\right)_p \tag{5}$$

The solvation of ions has been treated in numerous articles and
books (ref. 41).

The ligand-solvent interactions (Ib + IIb) have been investigated far less. Numerous solid adducts of organic molecules and crown
ethers have been isolated (ref. 42) even with acetonitrile (ref.
43), formamide (ref. 44) and nitromethane (ref. 45). However, quantitative data on these interactions are quite sparse (ref. 45, 46).

The electrostatic interaction energy between a donor atom and a
cation (Ic) is determined by ion-dipole, ion-quadrupole and ion-
induced dipole interactions. For the first term of this sume one
gets

$$\Delta H_{i-d} = - \frac{Ze\mu}{\varepsilon \, a^2} \tag{6}$$

Since dimethyl ether has a smaller dipole moment (1.30 D) than
water (1.85 D) or methanol (1.70 D) this term should disfavour complex formation with respect to the solvated state.

Another effect is also of importance. Repulsion between the solvent molecules forming the solvation shell destabilize the solvated
state (Id). This effect is easily demonstrated in the case of con-

secutive solvation steps of alkali ions in the gas phase, see
Table 2.1.

TABLE 2.1
Enthalpy changes for consecutive steps in the gas phase
(ΔH in kJ mol^{-1}) (from ref. 47).

n	1	2	3	4	5	6
Li$^+$	142	108	87	69	58	51
Na$^+$	100	83	66	58	51	45
K$^+$	75	67	55	49	45	42
Rb$^+$	67	57	51	47	44	
Cs$^+$	57	52	47	44		

It is expected that the linkage of binding sites in a suitable
arrangement in a single ligand supresses this destabilization
effect.

The changes for the cation interactions with solvent molecules
outside the first solvation shell (Ie) can easily be discussed with
reference to the Born equation (3) or the equation for the enthalpic contribution (4). The radius of the complexed cation a has to
be replaced by a new radius a' = a+s with the thickness of the
ligand s. Increasing shielding of the cation from the surrounding
medium may lead to a decrease in the stability constants.

Steric deformations of the ligand (If) become important for the
reaction of cations which are too small or too large for the intramolecular cavity.

During complex formation the ligand has to undergo several
changes in its conformation (IIc). In the final state some of its
flexibility is lost resulting in a negative term for the reaction
entropy. On the other hand, the most important contribution to the
reaction entropy is the translational entropy of the liberated solvent molecules (IId). In the case of the stronger solvated small
cations this factor should result in higher values for ΔS. The
increase in translational entropy at 25 $^\circ$C in water is given by:

$$T\Delta S_{tr} = 9.8 \cdot x \quad (\text{kJ mol}^{-1}) \tag{7}$$

x is the number of displaced solvent molecules minus one (ref. 48).

For the less solvated cations no solvent molecule may be liberated during complex formation (IIe). In this case a negative contribution to the overall entropy is expected. Two particles come together and form one particle as a result. Therefore, a loss of translational entropy is obvious.

In the following chapters experimental results will be used in the discussion of most of the theoretically mentioned factors on the complex stabilities. As far as possible a direct reference to them will be given.

3   EXPERIMENTAL METHODS

Before any conclusions can be drawn about a complexation reaction between a metal ion M^{n+} and a neutral ligand L

$$M^{n+} + L \rightleftharpoons ML^{n+}$$

the measurement of at least one actual concentration in the solution is necessary. Thus, using the mass balance equations it is possible to calculate the complex formation constant:

$$K = \frac{[ML^{n+}]}{[M^{n+}] \cdot [L]} \tag{8}$$

Most of the common experimental techniques have already been discussed in detail (ref. 49). Each method has its advantages and disadvantages. Nearly all physicochemical techniques fail if the stability constants are greater than 10^5 and an aprotic solvent is used.

To overcome these difficulties Schneider et al (ref. 50) performed disproportionation reactions with Ag^+. They monitored the concentration of free silver ions by means of a silver electrode.

$$ML^{n+} + Ag^+ \rightleftharpoons AgL^+ + M^{n+}$$

The ideal behaviour of this electrode was observed in all solvents. A similar competitive method is also described in existing literature using an Na^+ selective electrode (ref. 51).

As has already been mentioned it is not only desirable to know the stability constants but to obtain information about the reaction enthalpies and entropies. Often, log K is measured as a function of temperature. The reaction enthalpies are then obtained from

a van't Hoff plot. However, this method can be subject to serious
errors.

The calorimetric titration technique established by Izatt and
his co-workers (ref. 12, 52) is more straightforward. Stability
constants smaller than 10^5 can also be evaluated from the experimental data. Stronger formation constants can be obtained by means
of competitive titrations (ref. 53, 54).

In many cases, it is necessary to combine different techniques
in order to obtain the maximum information about a reaction.

4   SOLVENT EFFECTS UPON COMPLEX FORMATION

Due to different solvent properties the cation as well as the
ligand solvation may vary from one solvent to another. Therefore,
some selected values of dielectric constants are given.

TABLE 4.1
Dielectric constants of several solvents at 25 $^{\circ}$C (from ref. 56).

Solvent	ε
Water	78.39
Methanol	32.70
Ethanol	24.55
Acetonitrile	37.5
Pyridine	12.3
Dioxane	2.209

If the cation solvation is the main contribution to the changes
in the stability constants a linear dependence between log K and
$(1 - \frac{1}{\varepsilon})$ is expected from the Born equation. A plot of the observed
stability constants for the complexation of Ba^{2+} cations with
18-crown-6 in water-methanol and water-dioxane mixtures versus
$(1 - \frac{1}{\varepsilon})$ is shown in Figure 4.1.

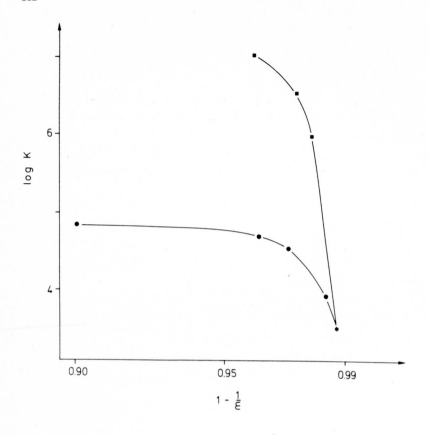

Fig. 4.1. Plot of log K versus $(1 - \frac{1}{\varepsilon})$ for the reaction of 18-crown-6 with Ba^{2+} in water-methanol (■) and water-dioxan (●) mixtures (from ref. 52,53,54,57 and 58)

In both solvent mixtures no linearity is obtained. The measured reaction enthalpies behave quite similarly. The values of the reaction entropies are nearly constant in all the solvent mixtures. Thus, mainly enthalpic factors are responsible for the increase in complex stability. Comparable results have also been reported from the complexation of other cations in solvent mixtures (ref. 19).

Quite different correlations are observed in the reaction of the macrobicyclic cryptands in solvent mixtures, see Fig. 4.2. The variation of log K with $(1 - \frac{1}{\varepsilon})$ for the K^{+}-cryptate complex differs significantly from that for crown ethers.

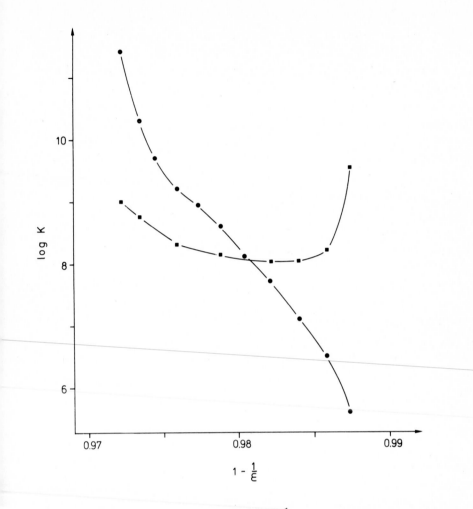

Fig. 4.2. Plot of log K versus $(1 - \frac{1}{\varepsilon})$ for the reaction of the cryptand (222) with Ag^+ (■) and K^+ (●) in water-acetonitrile mixtures (from ref. 56 and 58)

The reactions between the cryptand (222) and Ag^+ in water-acetonitrile mixtures give an example of the influence of cation selective solvation upon complex formation (ref. 59,60). This phenomenon is well known for some cations (ref. 61).

However, up to now, the influence of the solvent composition on the stability constants is not fully understood. More experimental data are required before there is a better understanding of this effect.

The importance of ligand solvation is evident from the studies of the enthalpies of transfer of 18-crown-6 and the cryptand (222) from water to methanol (ref. 62). From the measurements of the so-lution enthalpies of some crown ethers in methanol, only minor interactions between ligands and solvent molecules have been found (ref. 54). On several occasions a direct observation of the inter-actions between solvent molecules and ligands has been studied (ref. 46).

It is not possible to discuss the influence of the solvent on the complex formation with different ligands. However, if only the results obtained for reactions within the same solvent are compared, the influence of the solvent is constant. Therefore, the origin of the macrocyclic and cryptate effects may be determined from the results as obtained in only one solvent.

5   COMPLEX FORMATION WITH DIFFERENT NONCYCLIC LIGANDS WITH
     VARIOUS CATIONS

5.1   NONCYCLIC POLYETHERS

In the past, reports have only described a few examples of the reactions between noncyclic polyethers and alkali or alkaline-earth cations (ref. 13,26,27,63,64,65). Thermodynamic data are even more scarce.

Due to Frensdorff's hypothesis (ref. 13) that electrostatic re-pulsion between the terminal oxygens of small noncyclic ligands play an important role in their reactions, a direct comparison of the data concerning the reaction of noncyclic and macrocyclic li-gands possessing the same number of donor atoms may lead to incor-rect interpretations.

Recently more experimental results including thermodynamic data have been reported for the complexation of cations by polyethers (ref. 66) (see Figure 5.1.1) and by non-ionic surfactants (ref. 67) (see Figure 5.1.2).

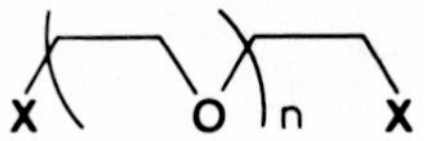

X = OH	n = 1	:	DEG
	n = 2	:	TEG
	n = 3	:	TeEG
	n = 4	:	PEG
	n = 5	:	HEG

X = OCH$_3$	n = 2	:	TG
	n = 3	:	TeG
	n = 4	:	PG
	n = 5	:	HG
	n = 6	:	HeG

Fig. 5.1.1. Different glycols and glymes

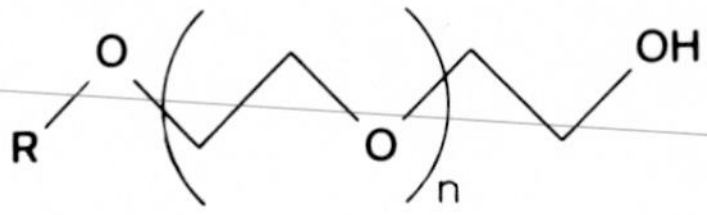

R=CH$_3$	n= 4 : PEGM
R=C$_{12}$H$_{25}$	n= 3 : B30
	n=22 : B35
R=C$_{16}$H$_{33}$	n=19 : B58
R=C$_{18}$H$_{35}$	n=19 : B99
R=C$_9$H$_{19}$⬡—	n=11 : JCO-720

Fig. 5.1.2. Nonionic surfactants

All known results for the complexation reaction between these
ligands and alkali and alkaline-earth cations in methanol solutions
are summarized in Table 5.1.1.

TABLE 5.1.1.

Log K (K in M^{-1}), ΔH (in kJM^{-1}) and TΔS (in kJM^{-1}) for the interaction of noncyclic polyethers with cations in methanol at 25 °C.

Ligand	n	Value	Na$^+$	K$^+$	Rb$^+$	Cs$^+$	Ca^{2+}	Sr^{2+}	Ba^{2+}
DEG	1	log K	$-$[a]	$-$[a]	$-$[a]	$-$[a]	$-$[a]	$-$[a]	$-$[a]
TEG	2	log K	$-$[a]	$-$[a]	$-$[a]	$-$[a]	$-$[a]	1.83[a]	3.40[a]
		$-$ΔH						13.8[a]	7.4[a]
		TΔS						$-$3.4[a]	11.9[a]
TG	2	log K	$-$[a]	$-$[a]	$-$[a]	$-$[a]	$-$[a]	$-$[a]	$-$[a]
TeEG	3	log K	$-$[a]	$-$[a]	$-$[a]	$-$[a]	$-$[a]	3.30[a]	3.65[a]
		$-$ΔH						10.8[a]	20.4[a]
		TΔS						8.0[a]	0.4[a]
TEG	3	log K	1.44[a] 1.28[b]	1.59[a] 1.72[b]	1.70[a]	1.43[a] 1.45[b]	$-$[a]	$-$[a]	2.15[a]
		$-$ΔH	15.1[a]	28.5[a]	23.4[a]	21.9[a]			12.4[a]
		TΔS	$-$6.9[a]	$-$19.5[a]	$-$13.7[a]	$-$13.8[a]			$-$0.2[a]
B30	3	log K	$-$[b]	3.76[b]	4.29[b]	3.88[b]	$-$[b]	3.84[b]	4.40[b]
		$-$ΔH		9.7[b]	9.2[b]	7.6[b]		3.6[b]	10.6[b]
		TΔS		11.7[b]	15.2[b]	14.4[b]		18.2[b]	14.4[b]

Ligand	n	Value	Na$^+$	K$^+$	Rb$^+$	Cs$^+$	Ca^{2+}	Sr^{2+}	Ba^{2+}
PEG	4	log K	-[a]	2.01[a], 2.05[d]	1.92[a]	1.29[a]	3.79[a]	3.58[a]	3.71[a], 3.96[d]
		-ΔH		24.9[a], 26.6[d]	24.7[a]	44.2[a]	3.7[a]	24.5[a]	28.4[a], 28.0[d]
		TΔS		-13.5[a]	-13.8[a]	-36.9[a]	17.8[a]	-4.2[a]	-7.3[a]
PG	4	log K	1.54[a], 1.47[b], 1.44[d], 1.52[e], 1.0[f]	2.07[a], 2.20[b], 2.27[d], 2.1[f]	1.98[a]	1.76[a], 1.85[b]	-	-	2.59[a], 2.51[d], 2.3[f]
		-ΔH	19.7[a], 16.8[d], 38.2[f]	44.7[a], 34.1[d], 36.4[f]	46.7[a]	38.6[a]			20.2[a], 23.6[d], 23.2[f]
		TΔS	-10.9[a]	-32.9[a]	-35.4[a]	-28.6[a]			-5.5[a]
PEGM	4	log K	-[c]	1.98[c]	1.77[c]	1.68[c]	-[c]	3.38[c]	3.35[c]
		-ΔH		43.2[c]	55.8[c]	35.1[c]		11.4[c]	27.0[c]
		TΔS		-31.9[c]	-45.7[c]	-25.6[c]		7.8[c]	-8.0[c]
HEG	5	log K	-[a]	3.05[a]	3.06[a]	1.67[a]	3.41[a]	4.28[a]	3.78[a]
		-ΔH		21.8[a]	20.1[a]	53.2[a]	4.4[a]	20.7[a]	34.4[a]
		TΔS		-4.5[a]	-2.7[a]	-43.7[a]	15.0[a]	3.6[a]	-12.9[a]

(continued)

Ligand	n	Value	Na+	K+	Rb+	Cs+	Ca2+	Sr2+	Ba2+
HG	5	log K	1.56[a] / 1.60[b]	2.55[a] / 2.55[b]	2.91[a] / 2.17[b]	2.10[a]	-[a]	-[a]	2.65[a]
		-ΔH	22.3[a]	39.0[a]	28.1[a]	41.1[a]			28.3[a]
		TΔS	-13.4[a]	-25.1[a]	-11.6[a]	-29.2[a]			-13.2[a]
HeG	6	log K	1.67[b]	2.87[b]		2.41[b]			
ICO-720	11	log K	1.52[c]	2.72[c]	2.77[c]	2.98[c]	-[c]	3.17[c]	3.68[c]
		-ΔH	36.8[c]	52.9[c]	55.3[c]	42.4[c]		13.7[c]	37.1[c]
		TΔS	-28.2[c]	-37.4[c]	-39.6[c]	-25.5[c]		4.3[c]	-16.2[c]
B58	19	log K	1.98[c]	2.59[c]	2.59[c]	2.42[c]	-[c]	2.51[c]	2.76[c]
		-ΔH	29.9[c]	79.3[c]	90.3[c]	91.4[c]		24.8[c]	60.3[c]
		TΔS	-18.6[c]	-64.6[c]	-75.6[c]	-77.6[c]		-10.5[c]	-44.6[c]
B99	19	log K	2.00[c]	2.62[c]	2.64[c]	2.47[c]	-[c]	2.66[c]	2.69[c]
		-ΔH	22.1[c]	72.6[c]	77.5[c]	85.7[c]		19.3[c]	65.3[c]
		TΔS	-10.7[c]	-57.7[c]	-62.5[c]	-71.7[c]		-4.2[c]	-50.0[c]
B 35	22	log K	2.07[c]	2.50[c]	2.56[c]	2.57[c]	2.48[c]	3.29[c]	2.57[c]
		-ΔH	26.8[c]	88.0[c]	88.8[c]	74.4[c]	1.5[c]	14.1[c]	76.1[c]
		TΔS	-15.0[c]	-73.8[c]	-74.3[c]	-59.8[c]	12.6[c]	4.6[c]	-61.5[c]

[a]from ref. 66
[b]from ref. 64
[c]from ref. 67
[d]from ref. 54
[e]from ref. 13
[f]from ref. 65

On comparison of the results for ligands with identical chain length and different end-groups, significant differences are obtained if the ligand is able to surround the complexed cation. The higher dipole moment of the hydroxyl end-groups compared with the ether groups leads to a stronger repulsion of the opposing parts of the ligand. Thus, the reaction enthalpies with glycols are lower than those for the glymes. The bulky end-groups hinder complex formation more than the small protons of the OH-group do. This effect is mainly responsible for the reaction entropies observed.

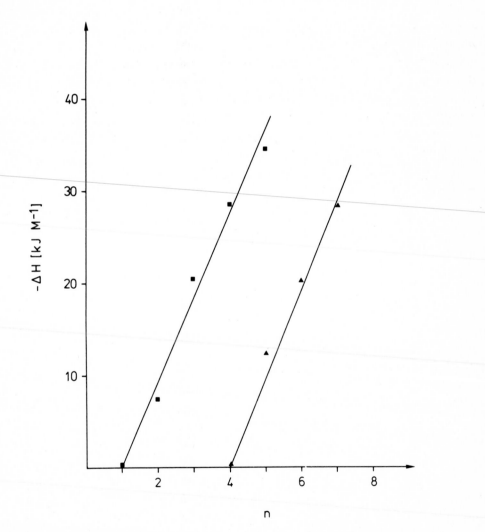

Fig. 5.1.3. The reaction enthalpies measured for the complexation of Ba^{2+} by glycols (■) and glymes (▲) in methanol as a function of the number of ether oxygen atoms in the ligand (from ref. 67).

The repulsion effect disappears if the oligoethylene glycol chain of the ligands reaches a specific length. Then, the values of the reaction enthalpies increase with the increasing number of donor atoms.

With the exception of Ba^{2+} no linear relationship is observed between the reaction enthalpies measured and the number of donor atoms (see Fig. 5.1.3).

From the gradient of both straight lines a medium bond strength of 9.1 ± 0.2 kJM^{-1} for a single Ba^{2+} - O bond can be calculated (ref. 67).

With regards to ligands, observations have been made of even longer oligoethylene glycol chains. Complexes with more than one cation complexed by one ligand molecule are observed (ref. 68, 69). This has to be taken into account in the evaluation of the results of the 1:1-complex formation from the experimental data. Under these conditions the reactions between different polyethylene oxides (see Fig. 5.1.4) and some cations are calculable. The results are summarized in Table 5.1.2.

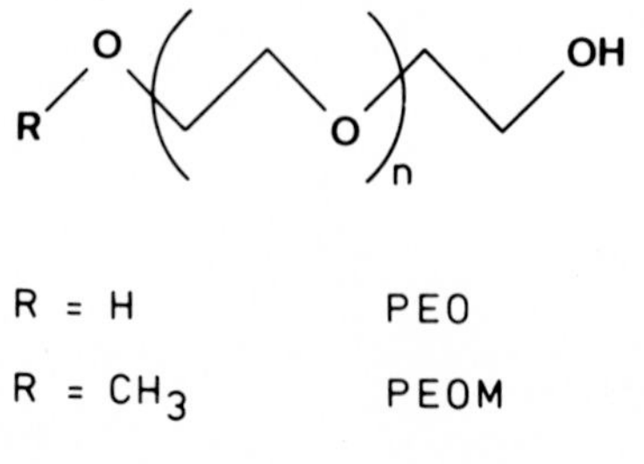

R = H   PEO

R = CH$_3$ PEOM

Fig. 5.1.4. Different polyethylene oxides

TABLE 5.1.2

Thermodynamic data for the 1:1-complex formation of polyethylene oxides with different cations in methanol solutions at 25 °C

Ligand	n	Value[a]	K^+	Rb^+	Cs^+	Ba^{2+}
PEOM 350[b]	7.2	log K	3.09	3.32	3.07	3.84
		$-\Delta H$	29.7	28.5	24.4	30.6
		$T\Delta S$	-12.1	-9.6	-7.0	-8.8
PEOM 750[b]	16.3	log K	2.52	2.67	2.58	2.63
		$-\Delta H$	75.6	74.2	69.2	68.8
		$T\Delta S$	-61.3	-59.0	-54.5	-53.9
PEOM 1900[b]	42.4	log K	2.48	2.59	2.51	2.56
		$-\Delta H$	112.8	94.2	102.1	82.1
		$T\Delta S$	-98.7	-79.5	-87.8	-67.6
PEO 3400[b]	76.8	log K	2.47	2.49	2.43	2.50
		$-\Delta H$	109.1	117.8	127.4	79.7
		$T\Delta S$	-95.1	-103.7	-113.6	-65.5
PEO 8000[b]	181.2	log K	2.47			2.54
		$-\Delta H$	101.9			77.9
		$T\Delta S$	-87.9			-63.5
PEO 14000[b]	317.4	log K	2.57			2.54
		$-\Delta H$	91.2			72.4
		$T\Delta S$	-76.6			-58.0
PEO[c]		log K	2.2			
		$-\Delta H$	112			
		$T\Delta S$	-72.2			

[a] K in M^{-1}, ΔH and $T\Delta S$ in kJM^{-1}
[b] The specification stands for a ligand with an average molecular weight indicated by the number, from ref. 69
[c] PEO with high molecular weight, from ref. 70

The results obtained with small polyethylene oxides are comparable with those for other small noncyclic polyethers. However, the reaction enthalpies measured rise again with an increasing number of donor atoms. In contrast to the smaller glycols, glymes and non-ionic surfactants the values of the reaction enthalpy which

were observed reach maximum values. In this case the cation is completely surrounded by the ligand. The repulsion between the terminal donor atoms does not affect complex formation. The steric requirements which lead to the ligands wrapping around the cations
are rather high. This is reflected in the reaction entropies. The
stability of the complexes formed is not very strong as a result
of this.

These ligands obviously substitute all the solvent molecules in
the immediate environment of the cations. Based on this assumption
it is possible, from the reaction enthalpies, to calculate a
'medium bond strength' between the complexed cations and the single
oxygen donor atom. The results are given in Table 5.1.3.

TABLE 5.1.3

Maximum reaction enthalpies ΔH_{max} (in kJM^{-1}) and solvation numbers
n_s for the calculation of bond energies between complexed cations
and one oxygen donor atom in methanol (from ref. 69).

	K^+	Rb^+	Cs^+	Ba^{2+}
$-\Delta H_{max}$ [a]	111 ± 3	106 ± 17	115 ± 18	81 ± 2
n_s	4	4	4	8
$-\dfrac{\Delta H_{max}}{n_s}$	27.8 ± 0.8	26.5 ± 4.3	28.8 ± 4.5	10.1 ± 0.3 9.1 ± 0.2[b]

[a] Mean value from the reaction of PEOM 1900 and PEO 3400.
[b] from ref. 67.

The 'individual bond strengths' calculated in this way may be
used to discuss the origin of the macrocyclic and cryptate effects.

A first test for the validity of these results is only possible
in the case of the Ba^{2+} ion. The calculated medium bond strength
from different experimental measurements between Ba^{2+} and one oxygen donor atom compare very well. However, one should always bear
in mind that these 'individual bond strengths' are only valid in
methanol solutions.

More data for the reaction between noncyclic polyethers and
other cations other than alkali and alkaline-earth ions are only
known for the complexation of Pb^{2+}, Ag^+ and Tl^+ in methanol solutions. From the values of the reaction enthalpy in Table 5.1.4 no

satisfactory relationship between the number of donor atoms and the measured reaction is obtained in the case of Pb^{2+}.

TABLE 5.1.4

Log K (K in M^{-1}), ΔH (in kJM^{-1}) and $T\Delta S$ (in kJM^{-1}) for the interaction of noncyclic polyethers with cations in methanol at 25 $^{\circ}C$.

Ligand	n	Value	Pb^{2+} [a]	Ag^{+} [b]	Tl^{+} [c]
TEG	2	log K	4.04		
		$-\Delta H$	2.9		
		$T\Delta S$	20.0		
TeEG	3	log K	3.17		
		$-\Delta H$	13.3		
		$T\Delta S$	4.8		
TeG	3	log K	2.06		1.57
		$-\Delta H$	7.2		
		$T\Delta S$	4.5		
PEG	4	log K	3.32		
		$-\Delta H$	31.4		
		$T\Delta S$	-12.5		
PG	4	log K	2.22	1.80	1.90
		$-\Delta H$	26.4	15.8	
		$T\Delta S$	-13.7	-5.6	
HEG	5	log K	3.61		
		$-\Delta H$	37.5		
		$T\Delta S$	-17.0		
HG	5	log K	2.22	1.82	2.30
		$-\Delta H$	38.9	23.0	
		$T\Delta S$	-26.2	-12.7	
HeG	6				2.55

[a] from ref. 71
[b] from ref. 72
[c] from ref. 64

The stability of complexes with noncyclic ligands always reaches a limiting value. A further enhancement is however possible if the ligands posses rigid end-groups. In this case enthalpic and entropic contributions favour the complexation reaction (ref. 73). The complexes are still less stable in comparison with similar crown ether complexes.

5.2 NONCYCLIC LIGANDS CONTAINING DIFFERENT DONOR ATOMS

Complexes with ligands containing nitrogen donor atoms have been investigated very intensively in the past (ref. 74). The chelate effect was discovered experimentally during these investigations (ref. 3). Polyamines are able to form complexes with almost all cations except alkali and alkaline-earth ions. All known data for the reactions in water have already been collected (ref. 75). Fewer data have been published for ligands containing sulphur as donor atoms (ref. 75).

Covalent contributions to the binding energies are expected in the interactions of both donor atoms with several cations as for example Ag^+, Pb^{2+}, Cu^{2+} and so on.

The 'individual bond strengths' between some cations and nitrogen and sulphur atoms are much more easily obtainable from experimental data when compared with the polyether ligands. Due to the high stability of the complexes formed it is even possible to directly study the interactions of one of these donor atoms with a complexed cation.

The enthalpies measured for the complex formation of different amines, diamines, polyamines, thioethers and other noncyclic ligands with mixed donor atoms with Ag^+ (refs. 72,76), Pb^{2+} (refs. 71,76), Cu^{2+} (ref. 77), Co^{2+} and Ni^{2+} (ref. 78) are used to calculate the bond strengths as summarized in Table 5.2.1.

TABLE 5.2.1

'Individual bond strengths' (in kJM^{-1}) for the interaction of different donor atoms with some cations in methanol solutions.

Donor atom	Ag^+	Pb^{2+}	Cu^{2+}	Co^{2+}	Ni^{2+}
O	6	3-8	0	0	0
NH	22	14	28	13	11
S	25	<1			

Up to now no further adaquate experimental data for other cations are obtainable from literature. In principle there is no fundamental difficulty in calculating the 'individual bond strengths' for more cations even those in other solvents.

6   MACROCYCLIC LIGANDS

In general macrocyclic ligands are able to form more stable complexes in comparison with their noncyclic analogues. However, their flexibility is partly reduced, therefore they are not able to accomodate every cation regardless of its size. The performed cavity of the macrocyclic ligands is important as far as the complexation behaviour of these ligands is concerned. The radii of macrocyclic polyether cavities and of some cations are given in Table 6.1.

TABLE 6.1
Cavity radii of macrocyclic polyethers and ionic radii of some cations (from ref. 79, 80).

Ligand	Radius (Å)	Cation	Radius (Å)	Cation	Radius (Å)
		Li^+	0.73	Ca^{2+}	1.00
12-crown-4	0.6	Na^+	1.02	Sr^{2+}	1.16
15-crown-5	0.9	K^+	1.38	Ba^{2+}	1.36
18-crown-6	1.4	Rb^+	1.49	Pb^{2+}	1.18
21-crown-7	1.9	Cs^+	1.70	Cu^{2+}	0.73
		Ag^+	1.15	Co^{2+}	0.57
				Ni^{2+}	0.69

There is a variation in the size of the cavity if other donor atoms are substituted for oxygens; however, these variations are small. Other factors dependent on the nature of the donor atoms are more important with regard to the stability of the complexes which are formed.

6.1   CROWN ETHER COMPLEXES

Crown ethers were the most frequently investigated monocyclic ligands in the past (ref. 19). These ligands are able to form strong complexes with alkali and alkaline-earth ions. They show distinct selectivity patterns towards the complexed cations, depending on the size of the cavity and the ionic radii.

Some of the most common unsubstituted crown ethers are shown in Fig. 6.1.1.

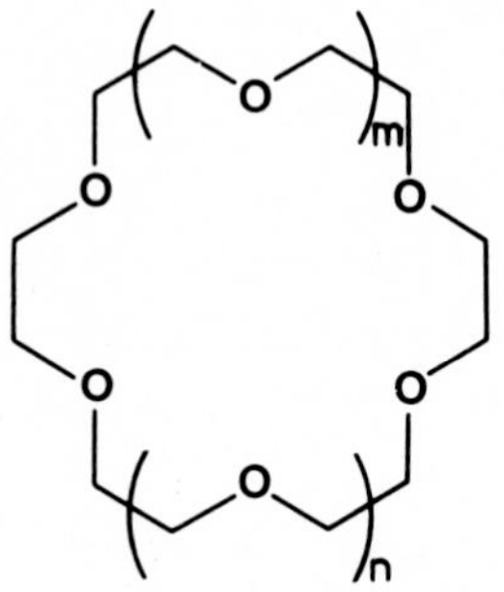

m = 0, n = 0 : 12C4
m = 0, n = 1 : 15C5
m = 1, n = 1 : 18C6
m = 1, n = 2 : 21C7

Figure 6.1.1. Some unsubstituted crown ethers

Due to their flexibility crown ethers are able to adopt different conformational forms in solution. Interaction with solvent molecules may stabilize one or even several conformations. Theoretical studies have been performed in order to obtain information about the lowest energetic structures (ref. 37,81). Experimental data from ultrasonic absorption of 18C6 in methanol give evidence of an isomeric relaxation process between two forms of the crown ether (ref. 82):

$$(18C6)_1 \rightleftharpoons (18C6)_2$$

$$\text{with } K_o = \frac{[(18C6)_2]}{[(18C6)_1]}$$

However, one form is predominant, because one gets $K_o \ll 1$ and $\Delta H_o = -10.9 \text{ kJM}^{-1}$. During the isomerization 1 mol of methanol is eliminated. This reaction happens before or at the time when a cation is complexed by a crown ether. Therefore, the procedure is identical for all cations in a given solvent and will not be given further consideration.

In all cases where the cavity diameter is smaller in comparison

with the diameter of the complexed cation the formation of complexes with more than one ligand molecule becomes possible. The structures of solid complexes with two ligand molecules and one cation
are well known (ref. 83). In Figure 6.1.2 a schematic drawing of
such a complex is shown.

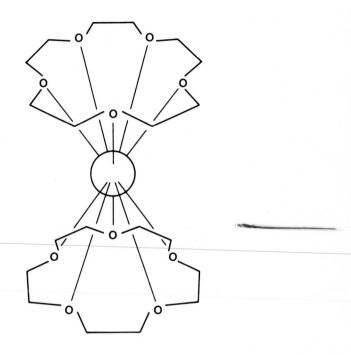

Fig. 6.1.2. Schematic drawing of a 2:1-complex (from ref. 84, by
kind permission)

If the cavity size of a crown ether is much larger than the radius of the cation a theoretical possibility exists that more than
one cation is complexed by the ligand. However, the electrostatic
repulsion works against the formation of complexes with such a
composition. Therefore, only a few examples of these complexes
have been reported (ref. 85). Up to now only the formation of 1:1-
and 2:1-complexes (ratio of ligand to cation) have been reported
in methanol solutions.

Some of the data published most recently are collected in
Table 6.1.1. More experimental results are easily obtained from
existing literature (ref. 19).

TABLE 6.1.1

Log K (K in M^{-1}), ΔH and $T\Delta S$ (in kJM^{-1}) for the reaction of unsubstituted crown ethers with alkali- and alkaline-earth cations in methanol at 25 °C.

Ligand	Value	Na$^+$	K$^+$	Rb$^+$	Cs$^+$	Ag$^+$	Ca^{2+}	Sr^{2+}	Ba^{2+}	Pb^{2+}
12C4	$\log K_1$	1.75[a]	1.55[a]			1.61[b]	2.53[a]	2.50[a]	2.56[a]	1.77[b]
	$-\Delta H_1$	8.4[a]	13.5[a]	9.9[a]	9.1[a]	10.7[b]	2.3[a]	12.4[a]	21.4[a]	13.9[b]
	$T\Delta S_1$	1.5[a]	-4.7[a]			-1.6[b]	12.1[a]	1.8[a]	-6.9[a]	-3.8[b]
	$\log K_2$	1.89[a]	1.54[a]	4.8[a]	5.9[a]	1.90[b]	4.3[a]	1.6[a]	2.38[c]	2.11[b]
	$-\Delta H_2$	21.5[a]	7.7[a]			27.9[b]			5.7[a]	9.6[b]
	$T\Delta S_2$	-10.8[a]	1.1[a]			-17.1[b]			7.8[c]	2.4[c]
15C5	$\log K_1$	3.42[d]	3.85[d]	4.07[d]	3.58[d]	3.65[d]	2.00[d]	3.20[d]	4.09[d]	3.92[e]
	$-\Delta H_1$	22.0[d]	31.0[d]	28.3[d]	21.2[d]	26.9[d]	9.3[d]	14.5[d]	20.9[d]	24.7[e]
	$T\Delta S_1$	-2.6[d]	-9.1[d]	-5.2[d]	-0.9[d]	-6.2[d]	2.1[d]	3.7[d]	2.3[d]	-2.4[e]
	$\log K_2$	2.77[d]	2.48[d]	2.47[d]	2.53[d]	3.07[f]		2.63[d]	2.61[d]	
	$-\Delta H_2$	9.7[d]	45.3[d]	44.0[d]	21.4[d]	7.2[f]		21.1[d]	38.8[d]	
	$T\Delta S_2$	6.0[d]	-31.2[d]	-30.0[d]	-7.0	10.2[f]		-6.2[d]	-24.0[d]	
18C6	$\log K_1$	4.32[d]	6.29[g]	5.82[g]	4.44[d]	4.58[d]	3.87[d]	6.84[g]	7.31[g]	6.99[b]
	$-\Delta H_1$	34.0[d]	54.9[d]	49.6[g]	49.9[d]	39.1[d]	11.2[d]	37.2[d]	48.5[d]	45.0[b]
	$T\Delta S_1$	-9.5[d]	-19.2[g]	-16.5[g]	-24.7[d]	-13.1[d]	10.8[d]	1.7[g]	-7.0[g]	-5.3[b]
	$\log K_2$	-[d]	-[d]	-[d]	4.14[d]		-[d]	-[d]	-[d]	
	$-\Delta H_2$				6.3[d]					
	$T\Delta S_2$				17.2[d]					

Ligand	Value	Na$^+$	K$^+$	Rb$^+$	Cs$^+$	Ag$^+$	Ca^{2+}	Sr^{2+}	Ba^{2+}	Pb^{2+}
21C7[h]	log K$_1$	1.73	4.22	4.86	5.01			1.77	5.44	
	$-\Delta H_1$	43.4	35.9	40.4	46.8			29.7	28.5	
	TΔS_1	-33.6	-11.9	-12.8	-18.3			-19.6	2.4	

[a] from ref. 86
[b] from ref. 76
[c] from ref. 87
[d] from ref. 84
[e] from ref. 71
[f] from ref. 88
[g] from ref. 55
[h] from ref. 53

All complexes formed with the ligand 12C4 are less stable when compared with the noncyclic analogues PEG and PG (see Table 5.1.1 and 5.1.4). The reduction of the complex stabilities is caused by lower values of the reaction enthalpies. Due to the fact that the donor atoms of 12C4 are already arranged in a fixed position they cannot achieve optimal interactions with the cation. The angle between the direction of the dipole and the charged cation is smaller than $180°$. The open-chain ligand does not have this difficulty. However, as has already been mentioned, in this case repulsion effects between the end groups of the noncyclic ligand may become important. The formation of 1:1-complexes with 12C4 are in nearly all cases favoured by the reaction entropies. Less steric changes of the cyclic ligand are possible during the reactions when compared with the noncyclic analogues.

The crown ether 15C5 is still too small to accommodate most ions. Only Na^+ and Ca^{2+} are almost the same size. Comparison with equivalent noncyclic ligands is impossible due to the important differences between HEG and HG.

Due to these findings 18C6 is the appropriate ligand in the discussion of the origin of the macrocyclic effect.

As a monocyclic ligand 18C6 is able to displace half of the four molecules surrounding an alkali ion. Thus, three donor atoms of 18C6 replace one solvent molecule from the inner solvation shell. Half the values of the reaction enthalpies measured for the complexation of alkali ions by 18C6 should therefore be comparable with the values given before in Table 5.1.3. Similar considerations have to be made in the case of the alkaline-earth cations. The results are summarized in Table 6.1.2.

TABLE 6.1.2
Reaction enthalpies of 18C6 and the maximum reaction enthalpies of noncyclic polyethers correlated with the number of replaced solvent molecules (in kJM^{-1}).

	Na^+	K^+	Rb^+	Cs^+	Ca^{2+}	Sr^{2+}	Ba^{2+}
n_s	4	4	4	4	8	8	8
$-\dfrac{2 \cdot \Delta H_{18C6}}{n_s}$	17.0	27.5	24.8	25.0	2.8	9.3	12.1
$-\dfrac{\Delta H_{max}}{n_s}$		27.8	26.5	28.8			10.1

Both sets of values agree very well with one another within the margin of experimental error. Thus, it is obvious that no important enthalpic contributions can be attributed to the origin of the macrocyclic effect in the case of the alkali and alkaline-earth cation.

The solvation of the doubly charged alkaline-earth cations is much stronger than the single charged alkali ions. Thus, the number of solvent molecules liberated during complexation is different. As a result the reaction entropies should favour the reaction of the divalent alkaline-earth cations. The experimental results confirm this expectation.

In the case of the Ag^+ and Pb^{2+} ions it is possible to calculate approximate reaction enthalpies from the values given in Table 5.2.1 for the reactions of these cations with 18C6. The experimentally obtained reaction enthalpies, see Table 6.1.1, do not differ from those obtained by calculation. However, the error in calculating the reaction enthalpy for the Pb^{2+} cation is largely due to uncertainty about the 'individual bond strength' with one oxygen donor atom.

These results clearly demonstrate the entropic origin of the macrocyclic effect.

The changes in ligand internal entropy are much lower in the case of the monocyclic ligands in comparison with the noncyclic polyethers. The noncyclic ligands have to adopt a coil structure during the reaction. Such a highly ordered conformation already exists in the macrocyclic ligands.

6.2 INFLUENCE OF STRUCTURAL CHANGES OF THE LIGAND UPON COMPLEX FORMATION

Ligands too small to accommodate a given cation form less stable complexes compared with 18C6. The donor atoms of the ligands are not able to achieve optimum interactions with the complexed cations. To compensate, many 15C5-ligands with different side-chains which also contain donor atoms have been synthesized (ref. 89). However, the only stability constants reported were those with Na^+ and K^+ in methanol solutions. When compared with 15C5 no increases in complex stabilities are observed for these ligands. Due to the fact that no thermodynamic data are available further explanation is not possible.

TABLE 6.2.1

Thermodynamic data for the complex formation between different 15C5 ligands and monovalent cations in methanol at 25 $^{\circ}$C (from ref. 84 and 90).

R	Value[a]	Na^+	K^+	Cs^+
-H	log K	3.42	3.85	3.58
	$-\Delta H$	22.0	31.0	21.2
	$T\Delta S$	-2.6	-9.1	-0.9
$-CH_2-O-\langle\textcircled{O}\rangle-OCH_3$	log K	2.90	3.17	2.63
	$-\Delta H$	22.6	33.8	32.6
	$T\Delta S$	-6.1	-15.8	-17.7
$-CH_2-O-\langle\textcircled{O}\rangle$ OCH_3	log K	3.24	3.32	
	$-\Delta H$	22.3	32.6	
	$T\Delta S$	-3.9	-13.7	

[a] K in M^{-1}, ΔH and $T\Delta S$ in kJM^{-1}

The only reported reaction enthalpies for the complexation of some alkali ions are for two 15C5 ligands with sidearms (ref. 90). The results are summarized in Table 6.2.1.

The measured reaction enthalpies indicate only in the case of Cs^+ the possibility of interactions between donor atoms of the sidearm and the cation. An unfavourable entropy change overcompensates the increase in the reaction enthalpy. As a result this complex is less stable when compared with 15C5. In the reaction of Na^+ and K^+ with the three different 15C5 ligands almost identical reaction enthalpies are observed.

Another means of increasing the stability of complexes with small crown ethers is to link two crown ethers in an appropriate way. Such a ligand molecule should be able to form a complex which is similar to the 2:1-complexes mentioned earlier. In this case, however, only one ligand molecule with two binding sites reacts with one cation. Several syntheses of such bis-crown ethers have been described in the literature (refs. 18,91). The interactions between all donor atoms of the bis-crown ether with the complexed cation depend upon the length of the linking moiety. In any case one expects that the complex formation with these ligands is favoured by entropic contributions when compared with the corresponding mono-crown ether. The known experimental results for different bis-(12-crown-4) ethers are given in Table 6.2.2.

TABLE 6.2.2

Thermodynamic data for the complexation of Na^+ and K^+ by bis-(12-crown-4) ethers and 12C4 in methanol at 25 $^\circ$C (from ref. 86 and 92)

X	Value[a]	Na^+	K^+
-	log K	1.59	1.72
$>C(C_2H_5)_2$	log K	3.25	1.82
$>C(CH_3)(C_{12}H_{25})$	log K	3.26	1.73
$-C_2H_4-$	log K	2.88	1.90
12C4	log K_1	1.75	1.55
	$-\Delta H_1$	8.4	13.5
	$T\Delta S_1$	1.5	-4.7
	log K_2	1.89	1.54
	$-\Delta H_2$	21.5	7.7
	$T\Delta S_2$	-10.8	1.1

[a] K in M^{-1}, ΔH and $T\Delta S$ in kJM^{-1}

No increase in complex stability over 12C4 is found for the shortest linking moiety. Thus, it is possible that only one crown ether part of the bis-crown ether is involved in the complex formation. Increasing the chain length between both crown ethers by one carbon atom nearly doubles the stability constant of the Na^+ complex. The complex stability of the bigger K^+ ion does not change. The addition of one further carbon atom into the chain increases the stability of this complex a little bit while the stability of the Na^+ complex is reduced. No detailed comment is possible because no further thermodynamic data have been published.

Similar behaviour is found in the case of several bis-(15-crown-5) ethers complexes with K^+ and Rb^+ (ref. 93). The values of the stability constants are only slightly smaller than the sum of log K_1 and log K_2 for the formation of 1:1- and 2:1-complexes with 15C5.

A more intensive study of the reactions with bis-crown ethers will give a better understanding of all factors influencing the complex formation.

The difference in complex stability between a bis-crown ether and the analogue crown ether has already been used to synthesize photoresponsive crown ethers (ref. 94). In response to photoirradiation the spacial distance between the two crown ether rings changes. This effect can be used very effectively for the transport of ions through liquid membranes.

Other examples of a different structural arrangement of both crown ether rings are also known (ref. 18). The rings are linked in such a way as to act as separate binding sites. Therefore, these ligands are able to complex two cations at the same time. As an example of this type of ligand spiro-bis-crown ethers can be considered (ref. 95). Their complexation reactions have been studied in pyridine solution. The results are summarized in Table 6.2.3.

The stability constants and reaction enthalpies for the complexation of the first cation by spiro-bis-crown ethers are nearly identical with the values found for the reaction of the quasi-analogue ligands 15C5 and 18C6. The reaction enthalpies are drastically reduced if a second cation is complexed by a spiro-bis-crown ether. This effect is caused by electrostatic repulsion between both cations. Same observations have been reported for two crown ether rings connected by a biaryl system (ref. 99).

Macrobicyclic crown ethers were used in a further attempt to increase the stability of the complexes formed. The structure of these ligands plays an important role in the complex stabilities (ref. 100). In the case of some macrobicyclic ligands higher stability constants have been found on comparison with the monocyclic crown ethers. Unfortunately no reaction enthalpies have been reported for the complexation of cations by these ligands.

TABLE 6.2.3
Thermodynamic data for the complexation of Na^+ by spiro-bis-crown ethers and related crown ethers in pyridine at 25 °C.

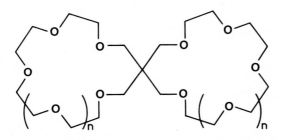

n	Value[a]	Na^+
1[b]	log K_1	2.80
	$-\Delta H_1$	9.5
	$T\Delta S_1$	6.4
	log K_2	1.44
	$-\Delta H_2$	≈0
2[b]	log K_1	3.10
	$-\Delta H_1$	39.3
	$T\Delta S_1$	-21.7
	log K_2	2.20
	$-\Delta H_2$	16.3
	$T\Delta S_2$	-3.8
15C5	log K	2.68[c]
	$-\Delta H$	15.1[d]
	$T\Delta S$	0.1
18C6	log K	>3[c]
	$-\Delta H$	36.2[d]

[a] K in M^{-1}, ΔH and $T\Delta S$ in kJM^{-1}
[b] from ref. 96
[c] from ref. 97
[d] from ref. 98

6.3   BENZO-CROWN ETHERS

This type of crown ether is the first example synthesized by Pederson (ref. 1). The electron density of the oxygen donor atoms is reduced due to the peripheral attachment of benzene rings to a crown ether. As a result the cation-dipole interactions are reduced and in comparison with unsubstituted crown ethers, smaller values of reaction enthalpies are expected. To help clarify this point, some results for different 18-crown-6 ligands substituted (see Figure 6.3.1) are summarized in Table 6.3.1.

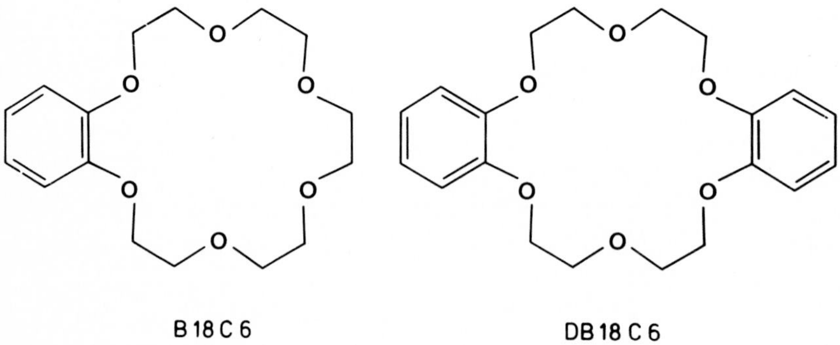

B 18 C 6   DB 18 C 6

Fig. 6.3.1. Different 18-crown-6 ethers substituted

As expected, the values of the reaction enthalpies measured decrease with an increasing number of benzene rings. In contrast, for a given cation the complex formation of DB18C6 are favoured by more positive TΔS terms than B18C6 or 18C6. Less conformational changes during the complex formation are possible due to to the rigidity of DB18C6.

TABLE 6.3.1

Log K (K in M^{-1}), ΔH (in kJM^{-1}) and TΔS (in kJM^{-1}) for the reaction of different 18-crown-6 ethers with alkali and alkaline-earth cations in methanol at 25 °C.

Ligand	Value	Na$^+$	K$^+$	Rb$^+$	Cs$^+$	Ca^{2+}	Sr^{2+}	Ba^{2+}
18C6[a]	log K	4.32	6.29	5.82	4.44	3.87	6.84	7.31
	-ΔH	34.0	54.9	49.6	49.9	11.2	37.2	48.5
	TΔS	-9.5	-19.2	-16.5	-24.7	10.8	1.7	-7.0
B18C6[b]	log K	4.21	5.29	4.48	3.95	2.28	5.12	5.48
	-ΔH	34.6	44.9	43.0	42.3	8.6	19.6	37.2
	TΔS	10.6	-14.6	-17.4	-19.7	4.4	9.6	-5.9
DB18C6[b]	log K	4.5	5.1	4.36	3.55			4.28
	-ΔH	31.2	40.0	28.6				21.2
	TΔS	-5.6	-10.9	-3.7				3.3

[a] values taken from Table 6.1.1
[b] from ref. 101

6.4 KETO-CROWN ETHERS

This class of crown ethers gained special interest due to their facile synthesis (ref. 102). Some mono- and disubstituted ligands are shown in Figure 6.4.1.

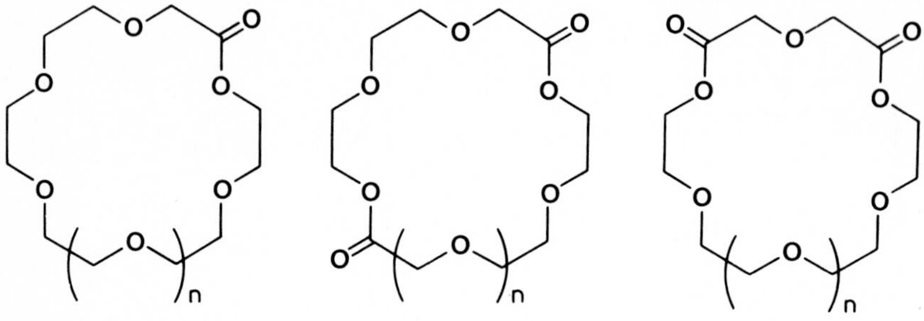

n = 0 : MK15C5
n = 1 : MK18C6

n = 1 : 2,11-DK18C6

n = 0 : 2,6-DK15C5
n = 1 : 2,6-DK18C6

Fig. 6.4.1. Diffferent mono- and diketo crown ethers

The data published are summarized in Table 6.4.1.

The complexes formed are less stable when compared with the unsubstituted crown ethers. Even the insertion of one keto group into the crown ether ligand causes a considerable drop in the measured complex stabilities. This effect is enlarged by the second keto group. Obviously the position of the second group is not important as it can be seen from the results for 2,6-DK18C6 and 2,11-DK18C6.

Compared with 18C6 the reaction enthalpies of 2,6-DK18C6 are drastically reduced with all cations. From the crystal structure of the K^+-complex of 2,6-DK18C6 it is known that the carbonyl oxygens do not participate in complex formation (ref. 104). Besides these findings one expects these oxygens to show an electron withdrawal effect from the neighbouring ether oxygens. The diketo-crown ethers also posses a more rigid conformation compared with unsubstituted crown ethers. As a result optimal interaction of the complexed cations with all donor atoms may not be achieved. These may be the factors responsible for the observed reaction enthalpies.

Also, the reaction entropies support these arguments. All $T\Delta S$-values of diketo-crown ethers are more than 10 kJM^{-1} more positive than the corresponding values for 18C6 at 25 °C.

TABLE 6.4.1

Stability constants and thermodynamic values for the reactions of monoketo- and diketo-crown ethers in methanol at 25 °C.

Ligand	Value[a]	Na$^+$	K$^+$	Rb$^+$	Cs$^+$	Ag$^+$
MK15C5[b]	log K	1.98	2.12			
2,6-DK15C5[b,c]	log K	–	–	–	–	–
MK18C6[b]	log K	3.27	4.18			
2,6-DK18C6[c]	log K	2.5	2.79	2.09	2.55	2.50
	$-\Delta H$	9.5	24.5	29.6	6.4	6.4
	$T\Delta S$	1.4	-2.6	-5.1	2.5	2.3
2,11-DK18C6[b]	log K	2.29	2.70			

[a] K in M^{-1}, ΔH and $T\Delta S$ in kJM^{-1}
[b] from ref. 103
[c] from ref. 104

The complex stabilities of diketo-crown ethers are of the same magnitude as measured for some noncyclic ligands (see Table 5.1.1). However, the values of the reaction enthalpy of diketo-crown ethers are much smaller. The importance of the entropy term on the stability of the complexes formed is obvious.

Using the general procedure for the synthesis of diketo-crown ethers it is possible to prepare ligands with different subcyclic units, for example benzene, furan, tetrahydrofuran and pyridine (ref. 102, 105, 106). One should note that all these different diketo-crown ethers form complexes which are less stable in comparison with unsubstituted crown ethers (ref. 106). Incorporation of a pyridine subcyclic unit into the ligand reduces the complex stability losses (ref. 107). The electron withdrawal effect of both carbonyl oxygens is reversed due to the pyridine aromatic ring.

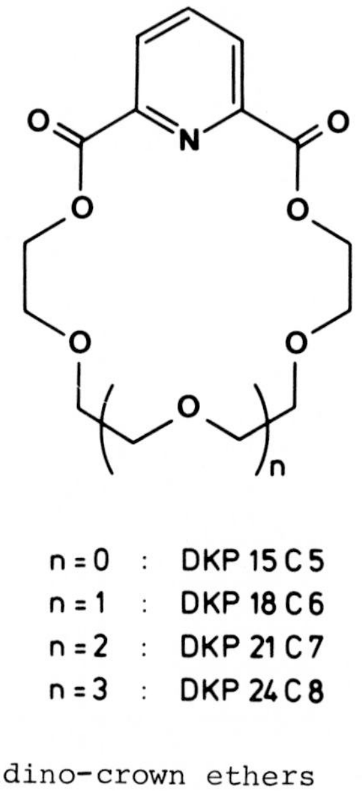

$$n = 0 \quad : \quad DKP\ 15\,C\,5$$
$$n = 1 \quad : \quad DKP\ 18\,C\,6$$
$$n = 2 \quad : \quad DKP\ 21\,C\,7$$
$$n = 3 \quad : \quad DKP\ 24\,C\,8$$

Fig. 6.4.2. Diketopyridino-crown ethers

This interpretation is supported by the reaction enthalpies measured. The values for diketopyridino-crown ethers are smaller than those for crown ethers but higher than those for diketo-crown ethers; on the other hand, nitrogen has a lower affinity than oxygen for alkali and alkaline-earth cations. In the case of Ag^+ the pyridino group is involved in the complex formation. Comparable reaction enthalpies with those of unsubstituted crown ethers are found.

TABLE 6.4.2

Stability constants and thermodynamic values for the reactions of diketopyridino-crown ethers in methanol at 25 °C.

Ligand	Value[a]	Na$^+$	K$^+$	Rb$^+$	Cs$^+$	Ca^{2+}	Sr^{2+}	Ba^{2+}	Ag$^+$
DKP15C5	log K	2.95[b]	2.52[b]	2.51[b]	2.41[b]	–[b]	2.48[b]	2.45[b]	2.56[c]
	–ΔH	8.4	29.1[b]	17.1[b]	9.3[b]		10.3[b]	25.6[b]	37.6[c]
	TΔS	8.4[b]	-14.8[b]	-2.8[b]	4.4[b]		3.8[b]	-11.7[b]	-23.1[c]
DKP18C6	log K	4.24[b] 4.29[d]	4.69[b] 4.66[d]	4.43[b] 4.24[e]	4.30[b]	–[b]	2.50[b]	4.34[b] 4.34[d]	5.00[b] 4.88[d]
	–ΔH	30.2[b] 25.9[d]	39.0[b] 38.9[d]	37.2[b] 38.0[e]	33.6[b]		11.1[b]	25.5[b] 25.2[d]	28.1[b] 32.7[d]
	TΔS	-6.1[b] -1.5[d]	-12.4[b] -12.4[d]	-12.0[b] -13.9[e]	-9.2[b]		3.1[b]	-0.8[b] -0.5[d]	-9.7[b] -5.0[d]
DKP21C7	log K	2.57[b]	3.60[b]	3.68[b]	3.76[b]	–[b]	2.75[b]	3.75[b]	
	–ΔH	36.9[b]	41.1[b]	44.8[b]	42.0[b]		10.5[b]	24.7[b]	
	TΔS	-22.3[b]	-20.6[b]	-23.9[b]	-20.6[b]		5.1[b]	-3.4[b]	
DKP24C8	log K	2.09[b]	2.82[b]	3.14[b]	3.41[b]	–[b]	2.55[b]	3.74[b]	
	–ΔH	32.7[b]	38.5[b]	38.9[b]	40.7[b]		18.1[b]	45.9[b]	
	TΔS	-20.8[b]	-22.5[b]	-21.1[b]	-21.3[b]		-3.6[b]	-24.6[b]	

[a] K in M^{-1}, ΔH and TΔS in kJM^{-1}
[b] from ref. 108
[c] from ref. 88
[d] from ref. 107
[e] from ref. 104

The complex formation with all diketopyridino-crown ethers is also favoured by entropic contributions due to the rigid structure of these ligands.

7 MACROCYCLIC AND MACROBICYCLIC LIGANDS WITH DIFFERENT DONOR ATOMS

Frensdorff observed a considerable drop in the stability of K^+ complexes with 18C6 and DB18C6 when one or two oxygen donor atoms were substituted by nitrogen or sulphur (ref. 13). The results were exactly the opposite in the reactions of Ag^+ with the substituted ligands. These first experimental results clearly demonstrate the influence of different donor atoms on complex formation. Thus, it is only possible to enhance the selectivity of a ligand for a special cation by having the optimal donor atoms available.

7.1 THIA CROWN ETHERS AND THEIR COMPLEXES

A systematic study of the influence of sulphur donor atoms on the complexation of noncyclic and macrocyclic ligands was carried out in water and water-methanol mixtures (ref. 109). From the experimental results the authors drew general conclusions about the replacement of oxygen donor atoms for sulphur atoms:

(a) no reactions between Na^+, K^+, Sr^{2+} and Ba^{2+} and any sulphur derivate of crown ethers are observed because the reaction enthalpies are too small.

(b) the stability of Ag^+ and Hg^{2+} complexes increases due to a more exothermic reaction.

(c) the replacement of oxygen by sulphur in the monocyclic ligands results in a significant decrease in log K caused by unfavourable $T\Delta S$ changes for Tl^+ and Pb^{2+}.

The 'individual bond strengths', given in Table 5.2.1, illustrate most of the experimental findings if there is no assumption of interactions between alkali and alkaline-earth cations and sulphur donor atoms.

Up to now, only a few experimental results have been published for the reactions of thia crown ethers, see Fig. 7.1.1, with different cations in methanol solutions. All results available are summarized in Table 7.1.1.

The interpretation of the experimental results obtained in methanol solutions is similar to that given for the results on water and water-methanol mixtures. The stability of complexes with alkali and alkaline-earth cations is reduced by the successive

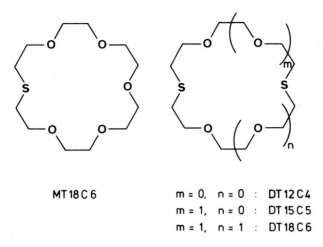

MT 18 C 6	m = 0, n = 0 :	DT 12 C 4
	m = 1, n = 0 :	DT 15 C 5
	m = 1, n = 1 :	DT 18 C 6

Fig. 7.1.1. Thia crown ethers

substitution of sulphur atoms for oxygen donor atoms. This is only caused by enthalpic contributions. The values of the reaction entropy found in the reaction of 18C6 and MT18C6 with these ions are nearly identical.

The reaction enthalpies for the complexation of Na^+ with MT18C6 and 15C5 are in excellent agreement. The Na^+ ion fits into the cavity of 15C5 well and is therefore able to achieve optimum interactions with all donor atoms. The same is possible with MT18C6. The number of donor atoms interacting with the complexed cation in both ligands are the same. The same is true for the remaining alkali ions and Ba^{2+}. However, these cations are too large to be encapsulated by 15C5 and therefore no maximum interactions between the cations and the donor atoms are possible. The cavity of MT18C6 is larger and the interactions are improved. As a result the values of the reaction enthalpies measured of MT18C6 are a little bit higher than those for 15C5.

The interactions between Pb^{2+} and sulphur donor atoms are negligible, see Table 5.2.1. The complex formed with DT18C6 is therefore less stable when compared with 18C6 due to less favourable enthalpic factors. The agreement between the reaction entropies for both ligands is very good.

Only in the case of the silver ion complexes, with different thia crown ethers, is an increase in the complex stability found when the number of sulphur donor atoms is increased. From the in-

TABLE 7.1.1

Log K (K in M^{-1}), ΔH (in kJM^{-1}) and $T\Delta S$ (in kJM^{-1}) for the reaction of thia crown ethers with different cations in methanol at 25 °C.

Ligand	Value	Na⁺	K⁺	Rb⁺	Ba²⁺	Ag⁺	Pb²⁺
DT12C4[a]	log K₁				–	7.56	4.01
	–ΔH₁					60.8	2.4
	TΔS₁					-17.8	20.4
	log K₂					5.29	1.78
	–ΔH₂					≈0	5.0
	TΔS₂					≈30	5.1
DT15C5[b]	log K					9.85	
	–ΔH					65.1	
	TΔS					-9.1	
MT18C6	log K	2.57[c]	3.61[c]	2.99[d]	3.4[c]	>5.5[c]	
	–ΔH	20.9[c]	37.7[c]	36.0[d]	25.5[c]	51.4[c]	
	TΔS	-6.2[c]	-17.1[c]	-18.9[d]	-6.2[c]		
DT18C6	log K		1.15[e]		–[a]	10.33[b]	4.76[a]
	–ΔH					64.0[b]	34.5[a]
	TΔS					-5.3[b]	-7.5[a]

[a] from ref. 76
[b] from ref. 88

[c] from ref. 110
[d] from ref. 111

[e] from ref. 13

dividual bond strengths given in Table 5.2.1 it is obvious that
stronger interactions are found between sulphur donor atoms in
comparison with oxygen donor atoms. The enhancement of the complex
stabilities is mainly due to enthalpic factors. Using the 'bond
strengths' given in Table 5.2.1 it is possible to calculate the
values of the reaction enthalpies. The agreement with the measured
reaction enthalpies is excellent. A lower value for the experimental reaction enthalpy is only found for DT18C6. The cavity of this
ligand is somewhat too big for the Ag^+ ion and a lower value may
result.

Due to the fact that all changes in the stability constant can
be reduced by changes in the reaction enthalpies, the macrocyclic
effect observed for these ligands is caused by entropic factors
only. This result is in accordance with that found for other monocyclic ligands.

7.2   AZA CROWN ETHERS AND CRYPTANDS

Both types of ligands are structurally very similar. Cryptands
may be characterized as macrobicyclic aza crown ethers. On the
other hand aza crown ethers are the monocyclic analogues of cryptands. Under these circumstances the results for the complexation
of cations by both ligand types should be discussed together.

From the NMR study of macrobicyclic diamines the existence of
three different conformations was deduced (ref. 15). The activation
energy of a conformational change, possibly a nitrogen inversion,
was found to be 32.2 kJM^{-1}. The different conformations were observed for the macrobicyclic cryptands also (ref. 112). Ultrasonic
absorption was used to study the cryptand (222) in aqueous solutions (ref. 113). The obtained experimental results were interpreted in terms of three ligand conformations in rapid equilibrium.

Crystal structures of cryptand complexes (cryptates) show that
the lone electron pairs of both nitrogen atoms are directed inside
the cavity (ref. 114). However, in solution the complexed ligand
is expected to exist in the same conformations as the free ligand
(ref. 115). This early statement of Lehn is supported now by the
experimental results for the kinetics of protonation and deprotonation (ref. 116). The dissociation reactions of many metal cryptands
are found to be acid catalysed (ref. 117). An interaction between
a proton and a nitrogen atom of the cryptand is only possible if
the electron pair is directed outside the cavity. All consideration
made for cryptates should also be valid for aza crown ethers. The

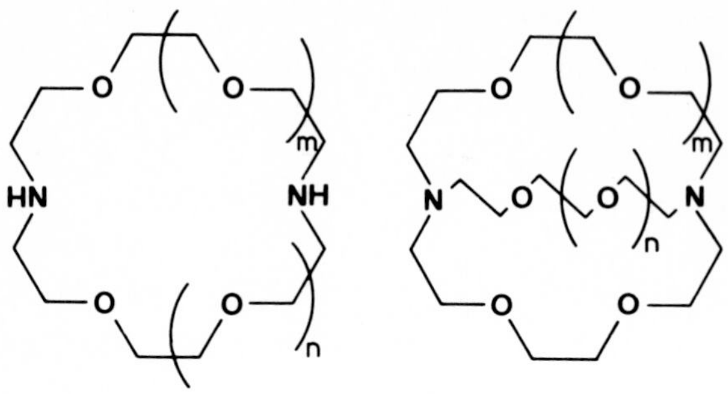

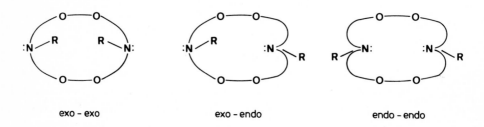

m = 1, n = 0 : 21   m = 1, n = 0 : 211
m = 1, n = 1 : 22   m = 1, n = 1 : 221
m = 2, n = 1 : 23   m = 2, n = 1 : 222

Fig. 7.2.1. Aza crown ethers and cryptands

various conformations possible for diaza crown ethers are shown
in Figure 7.2.2.

exo - exo   exo - endo   endo - endo

Fig. 7.2.2. Different conformational forms of incomplexed diaza
crown ethers (from ref. 72, by kind permission).

   If conformational changes of the macrocyclic and macrobicyclic
ligands take place during the complex formation the reaction
enthalpies measured are affected.
   The cavity size of the cryptands plays an important role in the
stability of the complexes formed. If the cavity is too large for
a complexed cation no optimal interactions between the donor atoms
and the cation can be achieved. On the other hand cations too big
for the cavity can only be accommodated if the ligand is deformed.
In both cases a reduction in the reaction enthalpy results. However, the reaction entropy is also affected.

TABLE 7.2.1

Cavity radii of cryptands (from ref. 118)

Ligand	Radius ($\overset{\circ}{A}$)
(211)	0.8
(221)	1.1
(222)	1.4

The cavity sizes of the different aza crown ethers are expected to be very similar to those of the analogue crown ethers. Together with some cation radii these values have already been summarized in Table 6.1.

Due to the high number of experimental data available for these ligands it is more convenient to discuss them in a separate chapter. The results of the complexation reactions will be treated separately for different types of cations in order to make them clearer.

8   COMPLEXES OF AZA CROWN ETHERS AND CRYPTANDS

8.1   ALKALI ION COMPLEXES

The introduction of one or two nitrogen atoms into macrocyclic crown ethers causes a decrease in the complex stabilities. Few experimental data for the reaction of monoaza and diaza crown ethers with alkali ions are available from publications. These results are summarized in Table 8.1.1.

Reaction enthalpies have only been published for the reactions of K^+ and Rb^+ with the diaza crown ether (22). Thus, a detailed discussion about the influence of nitrogen donor atoms on complex formation is not possible. The values of the reaction enthalpies are much lower when compared with the values of 18C6.

To enhance the complex stabilities monoaza ligands have been synthesized with an attached, conformationally mobile side chain to the nitrogen atom. The side chains carry donor atoms such as oxygen or nitrogen which are also expected to interact with the cation complexed in the crown ether cavity. The reactions of similar crown ethers with side chains have been discussed in Chapter 6.2. This class of ligands has been named "lariat ethers" by Gokel (ref. 122). Unfortunately in most publications only the stability constants of different "lariat ethers" with Na^+ or K^+ have been published (ref. 119, 120, 123, 124). More thermodynamic data have been reported for MA 15C5 and MA18C6 ligands containing an addi-

TABLE 8.1.1

Stability constants and thermodynamic data for the reactions of
monoaza and diaza crown ethers with alkali ions in methanol at
25 $^{\circ}$C.

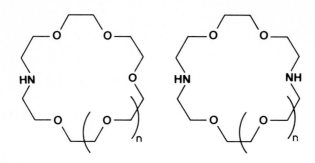

n = 0 : MA 15 C 5   n = 0 : 21
n = 1 : MA 18 C 6   n = 1   22

Ligand	Value[a]	Na$^+$	K$^+$	Rb$^+$	Cs$^+$
MA15C5	log K	2.06[b]	2.72[b]		
(21)	log K	_[d]	_[d]	_[d]	_[d]
MA18C6	log K	2.77[b]	4.18[b]		
		2.69[c]	3.90[e]		
(22)	log K	_[d]	2.04[e]	<1[d]	
			1.83[d]	1.2[f]	
	$-\Delta H$		4.7[d]		
	$T\Delta S$		5.7[d]		

[a] K in M^{-1}, ΔH and $T\Delta S$ in kJM^{-1}
[b] from ref. 119
[c] from ref. 120
[d] from ref. 121
[e] from ref. 13
[f] from ref. 118

tional ethylen glycol side chain with an increased number of oxygen
donor atoms (ref. 125). All available experimental results for the
substituted MA18C6 are given in Table 8.1.2.

TABLE 8.1.2

Stbility constants and thermodynamic data for the reaction of different substituted monoaza 18-crown-6 ethers in methanol at 25 °C.

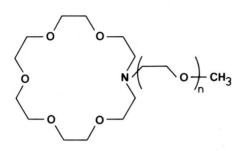

n	Value[a]	Na^+	K^+	Cs^+
0	log K	3.93^b		
1	log K	5.6^c 4.58^b	5.35^c	4.24^c
	$-\Delta H$	31.1^c	51.7^c	44.8^c
	$T\Delta S$	0.7^c	-21.3^c	-20.7^c
2	log K	5.7^c 4.33^b	5.5^c	4.34^c
	$-\Delta H$	28.0^c	52.4^c	49.3^c
	$T\Delta S$	4.4^c	-21.1^c	-24.6^c
3	log K	4.28^b 4.28^d	5.96^d	

[a] K in M^{-1}, ΔH and $T\Delta S$ in kJM^{-1}
[b] from ref. 123
[c] from ref. 125
[d] from ref. 119

The substitution of the nitrogen donor atom in the macrocyclic ring causes an important increase in the observed complex stabilities. The measured reaction enthalpies give no evidence of the participation of side chain donor atoms upon complex formation. With increasing chain length the values of the reaction enthalpy or the stability constants seem to be unaffected. Comparing the data in

Table 8.1.1 and 8.1.2, one finds that the substitution of the nitrogen atom proton by a methyl group already causes a big increase
in the complex stability. Other substituents lead to similar results. Therefore, the substitution itself is important and not the
special kind of substituent at the nitrogen atom.

The electron density of the nitrogen ring atoms does not even
play an important role in the stability of the complexes formed.
Another way of explaining the experimental results is that the
equilibrium of the different ligand conformers is disturbed. Bulky
substituents of the nitrogen atoms obviously shift the conformational equilibrium to the endo or, respectively, to the endo-endo conformation. Further conclusions can be drawn from the results obtained from the complexation reaction of different nitrogen substituted crown ethers, see Figure 8.1.3.

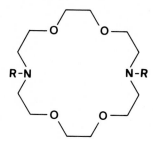

R = H   : 22
R = CH$_3$: 22 MM
R = (CH$_2$)$_9$CH$_3$: 22 DD

Fig. 8.1.3. Different substituted diaza crown ethers

These results are summarized in Table 8.1.3.

The substitution of both protons of the amino group of the
ligand (22) by methyl or long alkyl chains increase the observed
complex stabilities. This effect is only caused by enthalpic factors. The influence of the reaction entropy is exactly the opposite.

At this moment a more detailed discussion is not appropriate due
to the limited experimental results. However, in the next chapter
dealing with the complexation of alkaline-earth cations by aza
crown ethers this is possible.

The complex stabilities found for macrobicyclic ligands; i.e.
cryptands, are several orders of magnitude higher in comparison
with monocyclic and noncyclic ligands, see Figure 8.1.4.

TABLE 8.1.3

Stability constants (K in M^{-1}) and thermodynamic parameters (ΔH, $T\Delta S$ in kJM^{-1}) for the reaction of different stustituted diaza crown ethers with alkali ions in methanol at 25 $^{\circ}C$.

Ligand	Value	Li^+	Na^+	K^+	Rb^+	Cs^+
(22)[a]	log K	-	-	2.04	<1	-
				1.83		
	$-\Delta H$			4.7	<2	
	$T\Delta S$			5.7		
(22MM)[b]	log K	-	3.7	5.3	4.3	
(22DD)[c]	log K	-	3.02	4.00	3.51	3.08
	$-\Delta H$		16.8	31.5	34.4	21.7
	$T\Delta S$		0.4	-8.8	-14.5	-4.2

[a] values taken from Table 8.1.1
[b] from ref. 126
[c] from ref. 121

The cryptands form the most stable alkali complexes reported up to now. For the K^+ ion, which fits optimally into the cavity of the cryptand (222) and the crown ether 18C6, an increase in the measured stability constant of the cryptate complex by a factor of more than 10^4 is found on comparison with the crown ether complex. In relation to the noncyclic ligand PG an enhancement of the cryptand complex stability to the factor of 10^8 is observed.

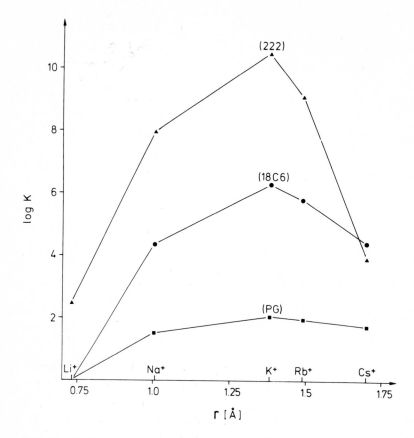

Fig. 8.1.4. Stability constants for the cryptand (222), the crown ether 18C6 and the noncyclic ligand PG with alkali ions in methanol at 25 °C (from ref. 55, 66 and 84).

The selectivity of a given cryptand against one cation depends very much on the ratio of the ligand cavity diameters to the size of the cation. This relation is demonstrated in Figure 8.1.5.

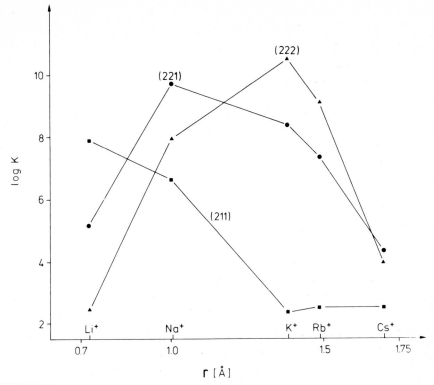

Fig. 8.1.5. Selectivity of some cryptands for the complexation of alkali ions in methanol at 25 °C (from ref. 122).

Some thermodynamic data for the reactions of the most common cryptands with alkali ions in methanol solutions are summarized in Table 8.1.4.

The reaction enthalpies measured reach a maximum value if the size of the complexed cations are equal to the cavity dimensions of the cryptand. If the cation is smaller the ligand has to change its conformation to minimize the distance between the complexed cation and the donor atoms. These conformational changes and the increased distances reduce the values of the measurable reaction enthalpies. If the cation is too large for a given cavity the ligand is deformed during the reaction. Even "exclusive" complexes are formed. Using cesium-133 NMR the existence of such complexes in different solvents for the system Cs^+ and (222) was confirmed by Popov et al (ref. 127). This results in a reduction of the reaction enthalpy.

TABLE 8.1.4

Stability constants (K in M^{-1}) and thermodynamic parameters (in kJM^{-1}) for the complexation of alkali ions by cryptands in methanol at 25 $^{\circ}C$ (from ref. 122).

Ligand	Value	Li^+	Na^+	K^+	Rb^+	Cs^+
(211)	log K	7.90	6.64	2.36	2.50	2.50
	$-\Delta H$	33.9	33.1	23.2	8.0	6.5
	$T\Delta S$	11.0	4.6	-9.8	6.2	7.7
(221)	log K	4.69	9.71	8.40	7.35	4.32
	$-\Delta H$	10.3	49.8	61.1	55.7	47.4
	$T\Delta S$	16.3	5.4	-13.4	-13.9	-22.9
(222)	log K	2.46	7.97	10.49	9.1	3.95
	$-\Delta H$	3.7	39.8	75.0	72.7	49.7
	$T\Delta S$	10.3	5.5	-15.4	-21.0	-27.3

For the cryptands (221) and (222) a specific trend in the values of $T\Delta S$ is observed. With increasing cation size they change from positive to negative values. As mentioned in Chapter 2, various factors may be responsible for these experimental results; however, the strength of the cation solvation decreases in the same order so less solvent molecules are liberated during complex formation. To summarize, various factors lead to the results observed and it is not possible to separate the contribution made by the various entropies.

It is possible to give an exact estimation of the energies required for the structural changes and deformations of the diaza crown ethers or the cryptands, both; these energetic contributions account for the differences between the reaction enthalpies (measured ΔH_m) of these ligands and the highest values for the reaction enthalpies, ΔH_c, calculated from the values of the reactions of polyethylene oxides. The values given in Table 6.1.2 may be used to calculate the difference between ΔH_m and ΔH_c. For the Na^+ cation the value extracted from the complexation reaction with 18C6 is used due to the fact that no direct reaction enthalpies with noncyclic ligands are measureable. The calculated deviations $\Delta = \Delta H_c - \Delta H_m$ are summarized in Table 8.1.5.

TABLE 8.1.5

Deviation Δ (kJM^{-1}) between calculated and reaction enthalpies measured for the complexation of alkali ions in methanol solutions at 25 °C.

Ligand	Na^+	K^+	Rb^+	Cs^+
(22)		-51		
(22DD)	-17	-24	-16	-36
(211)	-35	-88	-98	-108
(221)	-18	-50	-50	-68
(222)	-28	-36	-33	-65

The deviation Δ is nearly zero in the case of 18C6 (see Table 6.1.2). For the diaza crown ethers and cryptands strong negative values of Δ are found. They can be attributed to the conformational changes of the ligand. Due to the long alkyl chains at the nitrogen atoms of the ligand (22DD) the endo-endo conformation is preferred in the uncomplexed state. The ligand (22) has to adapt this conformation during complex formation. The calculated values of Δ clearly reflect this situation. They are much lower for (22DD) when compared with (22).

The molecular structures of cryptands and diaza crown ethers are related. Thus, similar behaviour of Δ can be expected. Also, all values of Δ calculated for cryptands are negative. They result mainly from the structural changes of the cryptands during complex formation. A distinct trend of the Δ values is evident. If the complexed cation is too big to be completely encapsulated by the cryptand, values of $\Delta \geq -50$ kJM^{-1} are calculated. For all cations with radii smaller than the cavity diameters, values of $\Delta < -40$ kJM^{-1} are obtained. Crystal structures reported by Weiss et al of sodium and potassium complexes with the cryptand (221) support this argument (ref. 128). The sodium complex was found to be an inclusive complex and the potassium an exclusive complex. The activation energy of a conformational change of macrobicyclic diamines was found to be 32.2 kJM^{-1} (ref. 15). This value is of the same order as found in the difference between ΔH_c and ΔH_m.

The observed stability constants for the reaction of diaza crown ethers and cryptands are therefore reduced due to enthalpic factors and favoured by entropic contributions with respect to the noncyclic ligands. Therefore it is possible to assign the cryptate effect mainly to entropic factors in the case of the alkali ions.

The possibility of the complexed cation interacting with solvent molecules outside the first solvation shell is reduced if benzene rings are introduced into the macrobicyclic cryptands. The thickness of such ligands, see Figure 8.1.6, is increased which reduces the interactions with further solvent molecules outside the cavity.

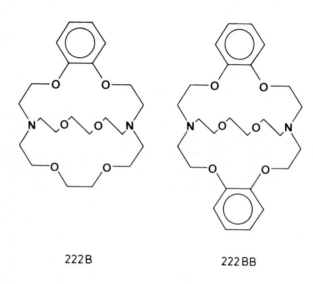

222B   222BB

Fig. 8.1.6. Benzo-substituted cryptands

The results published for both ligands in methanol solutions are given in Table 8.1.6.

The benzene rings reduce the cavity diameters, as the distance between the oxygen atoms at the benzene rings are smaller. The electro-negativity of the attached oxygen atoms is reduced and the rigidity of the ligand is increased. The possibility of optimal electrostatic interactions between the donor atoms and the complexed cation is therefore reduced.

The latter reasons are obviously responsible for the values measured. If the cation diameter is smaller when compared with the cavity diameter only a minor reduction in the stability constant and uneffected reaction enthalpies are observed.

The decrease in cavity size and the reduction in conformational flexibility leads to a much smaller measured stability constant for K^+, Rb^+ and Cs^+ complexes. As expected, both enthalpic and entropic contributions are responsible.

TABLE 8.1.6

Log K (K in M^{-1}), ΔH and $T\Delta S$ (in kJM^{-1}) for the reaction of benzo-substituted cryptands with alkali ions in methanol at 25 $^{\circ}$C.

Ligand	Value	Li^+	Na^+	K^+	Rb^+	Cs^+
(222)[a]	log K	2.46	7.97	10.49	9.1	3.95
	-ΔH	3.7	39.8	75.0	72.7	49.7
	$T\Delta S$	10.3	5.5	-15.4	-21.0	-27.3
(222B)[b]	log K	2.19	7.50	9.21	7.19	2.99
	-ΔH		39.7	65.3	57.7	31.8
	$T\Delta S$		2.9	-13.0	-16.8	-14.8
(222BB)[c]	log K	2.0	7.60	8.74	5.91	2.61
	-ΔH		42.5	66.2	53.7	38.5
	$T\Delta S$		0.7	-16.5	-20.1	-23.7

[a]from ref. 122   [b]from ref. 129   [c]from ref. 130

The introduction of the second benzene ring in (222BB) does not lead to lower values in the reaction enthalpy in comparison with the cryptand (222B) although the thickness of (222BB) increases. This result clearly demonstrates that the second solvation sphere of alkali ions is weak and does not play an important role in the complex formation with macrobicyclic ligands. X-ray diffraction studies of aqueous alkali halide solutions only show a very weak second solvation shell (ref. 131). In addition, it is found that the chemical shift of the lithium ion complexed by the cryptand (211) is essentially independent of the solvent (ref. 132). Therefore, no interactions between the complexed cation and additional solvent molecules take place.

The penultimate steps in the synthesis of diaza crown ethers and cryptands are the corresponding dilactam ligands (ref. 16), see Figure 8.1.7. Due to both carbonyl groups the nitrogen donor atoms are less basic, in comparison with the unsubstituted ligands. The rigidity of these ligands also increases. This therefore reduces the possibility of adapting different conformational forms.

No complexation of alkali ions by these dilactam ligands could be observed in methanol solutions (ref. 133). Mainly weak complexes with (222DL) are found (ref. 134) in other organic solvents. The thermodynamic data measured in acetonitrile solutions (ref. 133) give evidence that the complete loss of the cryptate effect is only caused by enthalpic factors.

n = 0 : 21 DL
n = 1 : 22 DL

n = 0, m = 0 : 211 DL
n = 1, m = 0 : 221 DL
n = 1, m = 1 : 222 DL

Fig. 8.1.7. Macrocyclic and macrobicyclic dilactam ligands

A variation in the number of oxygen donor atoms should also influence the complex stabilities. The experimental results for such a macrobicyclic ligand are given in Table 8.1.7.

TABLE 8.1.7
Stability constants for the complexation of alkali ions by a cryptand with four oxygen donor atoms in methanol at 25 $^\circ$C.

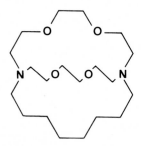

$22 C_8$

Ligand	Value[a]	K^+	Rb^+	Cs^+
(222)[b]	log K	10.49	9.1	3.95
$(22C_8)$[c]	log K	5.2	3.4	2.7

[a] K in M^{-1} [b] from ref. 122   [c] from ref. 126

Unfortunately no reaction enthalpies have been published. Owing to this, the influence of the different energetic contributions on complex stability cannot be discussed. However, one should notice that the K^+ complex with the macrobicyclic ligand ($22C_8$) is a factor of 10^3 more stable on comparison with the macrocyclic ligand (22). Both ligands possess the same number of donor atoms.

8.2   ALKALINE-EARTH ION COMPLEXES

For the reaction of alkaline-earth cations with different aza-crown ethers and cryptands in methanol solutions more experimental results have been published (ref. 19). Some of the most recent data are given in Table 8.2.1.

The most important energetic contribution for the complexation reactions of the ligands (21), (22) and (23) comes from the reaction entropy.

In some reactions even positive values of the reaction enthalpies are observed. According to the equation (2) a negative value of ΔG and in turn a positive value for the stability constant is only possible if the value of ΔS overcompensates that of ΔH.

Substitution of methyl (22MM) or decyl groups (22DD) for the protons of both amino groups in ligand (22) considerably increase the values of the reaction enthalpy. However, the enthalpies measured for the complexation reactions with (22MM) and (22DD) are almost identical. The voluminous groups attached to the nitrogen atoms favour the endo-endo configuration of the uncomplexed diaza crown ethers, see Figure 7.2.2. Ligands without these substitutents adapt this conformation during the complex formation process. On the other hand the substituents make complex formation more difficult. Consequently, the reaction entropies decrease going from (22) to (22MM) and further for (22DD). Changes in ligand solvation due to different nitrogen substituents are also possible, these influence the values measured.

Some experimental results for the reaction of different mono-aza crown ethers with a side chain have been reported (ref. 125). These data for the Ca^{2+} cation give no evidence of the interactions of these side chains containing additional donor atoms with the complexed cation.

TABLE 8.2.1

Stability constants and thermodynamic values for the complexation of alkaline-earth cations by different diaza crown ethers in methanol at 25 °C.

Ligand	Value[a]	Ca^{2+}	Sr^{2+}	Ba^{2+}
(21)[b]	log K	2.56	3.14	2.72
	$-\Delta H$	4.3	-10.3	<0
	$T\Delta S$	10.2	28.1	
(22)[b]	log K	3.87	5.99	6.12
	$-\Delta H$	-5.6	9.0	10.0
	$T\Delta S$	27.6	25.0	24.7
(22MM)[c]	log K	4.2	6.5	6.9
	$-\Delta H$	15.6	25.0	34.6
	$T\Delta S$	8.3	11.9	4.6
(22DD)[b]	log K	2.51	5.44	5.84
	$-\Delta H$	11.7	23.2	32.9
	$T\Delta S$	2.6	7.7	0.3
(23)[b]	log K	1.86	3.58	5.39
	$-\Delta H$	0	-7.3	8.5
	$T\Delta S$	10.6	27.6	22.1

[a] K in M^{-1}, ΔH and $T\Delta S$ in kJM^{-1}
[b] from ref. 135
[c] from ref. 136

The macrobicyclic ligands are able to form complexes which are much more stable than monocyclic ligands. Experimental results are given in Table 8.2.2.

TABLE 8.2.2

Stability constants and thermodynamic parameters for the complexation of alkaline-earth cations by cryptands in methanol at 25 $^{\circ}$C.

Ligand	Value[a]	Ca^{2+}	Sr^{2+}	Ba^{2+}
(211)	log K	5.45[b]	2.50[b]	2.53[b]
	-ΔH	2.4[b]	0.2[b]	5.5[b]
	TΔS	28.6[b]	14.0[b]	8.9[b]
(221)	log K	9.92[c]	11.04[c]	10.4[b]
	-ΔH	32.5[b]	43.0[b]	38.2[b]
	TΔS	23.9[b]	19.7[b]	20.9[b]
(222)	log K	8.16[b]	11.75[c]	12.9[d]
	-ΔH	22.0[b]	42.5[b]	68.9[b]
	TΔS	24.4[b]	24.3[b]	4.4[b]

[a] K in M^{-1}, ΔH and TΔS in kJM^{-1}
[b] from ref. 135
[c] from ref. 137
[d] from ref. 138

Alkaline-earth cations too big to fit into the ligand cavity are only able to form exclusive complexes. The observed stability constants reach the highest values if the ligand cavity and the size of the cation are nearly the same. This was discussed in detail in the previous chapter.

Some differences between the reactions of cryptands with alkali and alkaline-earth cations are worthwhile noticing. The values of the reaction entropy for the complexation of the divalent cations are positive in all cases. For the monovalent ions they change from positive to negative values. The differences in solvation between alkali and alkaline-earth cations is the most probable explanation for these observations since no changes in ligand solvation occur. However, due to the different sizes of the cations changes in the ligand structure occur, achieving optimal interactions between the complexed cation and the donor atoms of the ligand.

Fortunately, the K^{+} and Ba^{2+} ions are almost identical in size (ref. 80) thus conformational changes of the ligands during the complex formation should be the same. Only the cryptand (222) is able to encapsulate both ions complete without deformation. In the case of the smaller cryptands, difficulties arise owing to the formation of exclusive complexes. The results obtained for the reaction of the ligand (222) with K^+ and Ba^{2+} in methanol solutions is the basis for further discussion.

The values of the reaction enthalpies of both cations are expected to be similar. The process of desolvating these cations and establishing new interactions with the donor atoms of the ligand should be independent of the charge of the cation if only ion-dipole interactions are involved. The reaction enthalpies obtained experimentally are quite similar. Under these conditions entropic factors are responsible for the difference in both stability constants. In order to eliminate the influence of ligand changes during the reaction, the difference of the $T\Delta S$-values for both cations can only be considered. From Tables 8.1.4 and 8.2.2 one gets for the difference at 25 ^{o}C:

$$T\Delta S_{Ba^{2+}} - T\Delta S_{K^+} = 19.9 \ [kJM^{-1}]$$

The entropy of fusion for methanol is estimated to be 18.0 JK^{-1} M^{-1} (ref. 139). Using both values it is possible to calculate the difference in the number of liberated solvent molecules during complex formation of both cations. This simple model indicates that a further four solvent molecules are set free during the complexation of Ba^{2+} compared with K^+. This result is in accordance with the difference in the solvation number of both cations, see Table 6.1.2.

Equation (7) may be used to calculate the increase in translational entropy for the complex formation of K^+ and Ba^{2+}. The difference between both $T\Delta S$ values is nearly double that of the experimental result even after corrections for methanol as a solvent are made. Due to the very complex reactions, deviations from the numerical value in equation (7) are expected (ref. 48). A further means of testing the validity of the result obtained from the simple model is to use the values of $T\Delta S$ for the reaction of the monocyclic ligand 18C6 with both cations (Table 6.1.1) in the same way. The number of solvent molecules liberated during complex formation is found to differ by two as expected. The influence of the reaction entropy upon the complex stabilities of alkali and alkaline-earth cations is clearly demonstrated. This effect however should diminish in solvents with lower solvating abilities.

A discussion of the enthalpic contributions to the complex formation of diaza crown ethers and cryptands is possible only for the Ba^{2+} ion. For this cation an 'individual bond strength' with oxygen donor atom in methanol is known, see Table 6.1.2. This 'bond strength' is used to calculate the values of the reaction enthalpy for the different ligands. The difference between calculated and reaction enthalpies measured are shown in Figure 8.2.1.

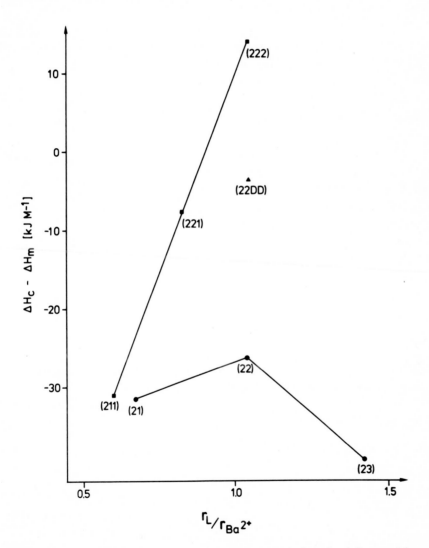

Fig. 8.2.1. Difference between calculated (ΔH_c) and reaction enthalpies measured (ΔH_m) as a function of the ligand-to-barium radii.

The reaction enthalpies measured with the macrocyclic ligands (21), (22) and (23) are approximately 30 kJM^{-1} more negative than the calculated ones. This difference is in accordance with the observations of the alkali ions. Conformational changes of the ligands account for this observation. The difference is close to zero for the ligand (22DD) due to the preference of one conformational form of the uncomplexed ligand.

For the macrobicyclic ligands varied results are found. With increasing cavity diameters the difference between calculated and reaction enthalpies measured reaches even positive values (ref. 67). Thus, the energy for the nitrogen inversion is still taken away from the reaction enthalpies. The spatial arrangements of all donor atoms in the ligand (222) is obviously better than for the non-cyclic ligands. This may account for the positive difference between calculated and measured reaction enthalpy. The complexation of Ba^{2+} by cryptands is an example of measureable positive contributions of reaction enthalpy and entropy to the observed complex stability.

X-ray diffraction studies of aqueous alkaline-earth chloride solutions give experimental evidence of the existence of a second solvation shell (ref. 140). Therefore, big differences in the complex formation between the ligands (222), 222B) and (222BB) are anticipated, see Table 8.2.3. Interactions between complexed divalent cations and the solvent are concluded from the experimental volumes of complexation (ref. 141). Further evidence can be drawn from the solid-state structure of barium complex with (222) (ref. 142). Compared with the cryptand (222) the values of the reaction enthalpy decreases with an increasing number of benzo substituents. The complexed cations are shielded more and more from the surrounding solvent. The reaction entropies remain nearly constant. Thus, the reduction of the complex stabilities of the ligands (222B) and (222BB) is only caused by enthalpic changes.

TABLE 8.2.3

Stability constants and thermodynamic values for the complexation of alkaline-earth cations by different benzo substituted cryptands in methanol at 25 $^\circ$C.

Ligand	Value[a]	Ca^{2+}	Sr^{2+}	Ba^{2+}
(222)[b]	log K	8.16	11.75	12.9
	$-\Delta H$	22.0	42.5	68.9
	$T\Delta S$	24.4	24.3	4.4
(222B)	log K	7.19[c]	10.52[c]	11.05[c]
	$-\Delta H$	17.7[d]	34.3[d]	53.9[d]
	$T\Delta S$	23.2	25.5	8.9
(222BB)	log K	5.94[c]	9.05[c]	8.85[c]
	$-\Delta H$	6.4[d]	25.9[d]	33.5[d]
	$T\Delta S$	27.4	25.5	16.8

[a] K in M^{-1}, ΔH and $T\Delta S$ in kJM^{-1}
[b] from Table 8.2.2
[c] from ref. 143
[d] from ref. 144

Some of the macrocyclic and macrobicyclic dilactam ligands, shown in Figure 8.1.7, form complexes with alkaline-earth cations. No reactions with the diaza crown ether analogues are observable. The measured data are given in Table 8.2.4.

The stability of the complexes formed is reduced by a factor of 10^5 up to 10^{10} compared with the unsubstituted ligands. Both enthalpic and entropic factors are responsible. The complexes of the dilactam cryptands are even less stable than most crown ether complexes. This example clearly shows how 'little' structural changes drastically influence the complexing properties of a ligand.

TABLE 8.2.4

Stability constants and thermodynamic parameters for the complexation of alkaline-earth cations by the dilactam derivatives of diaza crown ethers and cryptands in methanol at 25 $^{\circ}$C (from ref. 133).

Ligand	Value[a]	Ca^{2+}	Sr^{2+}	Ba^{2+}
(22DL)	log K	-	-	-
(211DL)	log K	-	-	-
(221DL)	log K	3.49	2.98	-
	$-\Delta H$	6.0	2.4	
	$T\Delta S$	13.8	14.5	
(222DL)	log K	3.58	4.01	2.90
	$-\Delta H$	18.8	11.8	14.5
	$T\Delta S$	1.5	11.0	2.0

[a] K in M^{-1}, ΔH and $T\Delta S$ in kJM^{-1}

8.3 SILVER ION COMPLEXES

The interactions of silver ions with nitrogen and sulphur donor atoms are much stronger than with oxygen donor atoms, see Tab. 5.2.1 The bonds between Ag^+ and nitrogen or sulphur atoms possess covalent character whilst pure electrostatic interactions take place with oxygen atoms.

The experimental results for the most common macrocyclic and macrobicyclic ligands are summarized in Table 8.3.1.
From the results for the noncyclic ligand DAOO, the macrocyclic ligand (22) and the macrobicyclic ligand (222) one can easily conclude the absence of a macrocyclic and cryptate effect. Same observations are reported for the reaction of these three ligands in water (ref. 30).

With the exception of (22DD) the values of the reaction enthalpy for macrocyclic ligands are smaller than for the noncyclic ligand DAOO. With an increasing number of donor atoms no enhancement in the values of the reaction enthalpy occurs. As already discussed in the previous chapters structural changes of the ligand during the reaction (e.g. the inversion of the nitrogen donor atoms, the orientation of oxygen donor atoms inside the cavity and so on) are

Table 8.3.1

Stability constants and thermodynamics parameters for the reaction of Ag^+ with mono- and bicyclic ligands in methanol at 25 $^{\circ}C$.

Ligand	log K[a]	$-\Delta H$[b]	$T\Delta S$[b]
DAOO[c,d]	9.59	58.3	-3.8
(21)[d]	7.63	34.6	8.8
(22)[d]	10.02	44.9	12.0
(22DD)[d]	10.28	61.1	-2.7
(23)[d]	9.60	53.4	1.1
(211)[d]	10.46	54.6	4.8
(221)[d]	14.44	81.9	0.1
(222)[d]	12.22	68.3	1.1
(222B)[e]	11.98	65.1	3.0
(222BB)[e]	11.84	65.4	1.9

[a] K in M^{-1}
[b] ΔH and $T\Delta S$ in kJM^{-1}
[c] $H_2N-CH_2-CH_2-O-CH_2-CH_2-O-CH_2-CH_2-NH_2$, 1,8-diamino-3,6-dioxaoctane (DAAOO)
[d] from ref. 72
[e] from ref. 88

mainly responsible for these findings. Also for the Ag^+ cation this explanation is in accordance with the results obtained for the ligand (22DD).

The strongest complex with a macrobicyclic ligand is formed with the cryptand (221). Its cavity dimensions are optimal for the accommodation of Ag^+. Using the values for the 'individual bond strength' from Table 5.2.1 it is possible to compare the calculated with the experimental reaction enthalpies. The difference is shown in Figure 8.3.1.

These results are very similar to those for Ba^{2+}. As a result, the interpretation is also the same.

The cryptands (222B) and (222BB) and the ligand (222) show identical behaviour. Thus, interactions between the complexed Ag^+ and further solvent molecules in a second solvation shell can be excluded. This is in agreement with the results obtained from aqueous silver perchlorate solutions by X-ray diffraction studies (ref. 145).

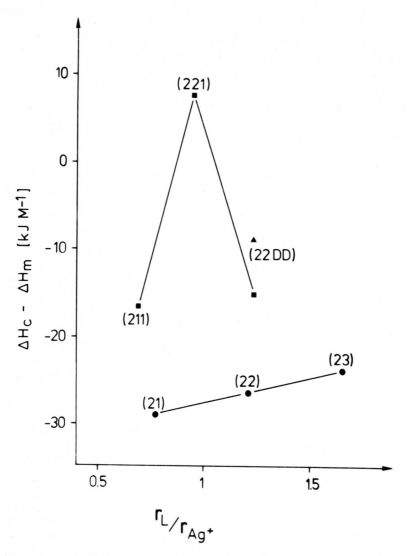

Fig. 8.3.1. Difference between calculated (ΔH_c) and reaction
enthalpies measured (ΔH_m) as a function of the ratio of the
ligand-to-silver radii.

The macrocyclic and macrobicyclic dilactam ligands, shown in
Figure 8.1.7, do not form any detectable complex with Ag^+ in methanol solution (ref. 133). The basicity of the nitrogen donor atoms
is strongly reduced. Thus no interactions with the complexed Ag^+
are possible.

Just recently the preparation of some diaza crown ethers and one cryptand with oxygen and sulphur donor atoms has been reported together with their complexing properties towards Ag^+ in many different solvents (ref. 146). For the purpose of this article only use of the results in methanol solution is made, Table 8.3.2.

Table 8.3.2
Stability constants and thermodynamic parameters for the reaction of Ag^+ with diaza crown ethers and cryptands with different donor atoms in methanol at 25 $^\circ$C (from ref. 146).

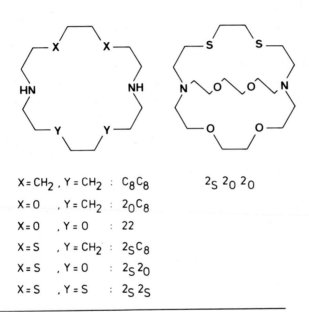

$X = CH_2$, $Y = CH_2$: C_8C_8

$X = O$, $Y = CH_2$: 2_OC_8

$X = O$, $Y = O$: 22

$X = S$, $Y = CH_2$: 2_SC_8

$X = S$, $Y = O$: 2_S2_O

$X = S$, $Y = S$: 2_S2_S

$2_S 2_O 2_O$

Ligand	log K^a	$-\Delta H^b$	$T\Delta S^b$
DAO^c	6.93	47.2	−7.8
(C_8C_8)	4.8	48.1	−20.8
(2_OC_8)	9.0	50.6	1.0
(22)	10.02	51.4	5.7
(2_SC_8)	10.9	−	−
(2_S2_O)	11.5	67.7	−2.0
(2_S2_S)	13.7	83.2	−5.0
$(2_S2_O2_O)$	13.4	93.2	−16.7

^{a}K in M^{-1}
$^b\Delta H$ and $T\Delta S$ in kJM^{-1}
$^cH_2N(CH_2)_8NH_2$, 1,8-diaminooctane (DAO), from ref. 72

The complex stability with the macrocyclic diamine (C_8C_8) is smaller when compared with the noncyclic analogue DAO. This reduction is caused by entropic factors only. Both reaction enthalpies are identical. Thus, the flexibility of the nitrogen atoms in (C_8C_8) is very high. In this ligand the energy for structural changes is the same as for the noncyclic ligand. With an increasing number of oxygen donor atoms the reaction enthalpies measured almost remain constant. Obviously the energy for structural changes of the ligand during complex formation increases. This is not surprising because the neighbouring donor atoms with their own dipole moments generate a repulsion effect for the nitrogen atom during inversion. In contrast, the values of the reaction entropies change considerably with an increasing number of oxygen donor atoms. Due to the hydrophobic behaviour of the alkyl chains they are in direct contact with one another. The cavity for the Ag^+ ion has to be formed during complexation which causes the large negative value for the reaction entropy of the ligand (C_8C_8). The same explanation is valid for the results of diaza crown ethers with sulphur donor atoms. The difference in the 'individual bond strength' between sulphur and oxygen is responsible for the changes of the reaction enthalpies.

The cavity of the macrocyclic and macrobicyclic ligands are enlarged because the van der Waals' radius of sulphur ($r = 1.85$ Å) is larger than that of nitrogen ($r = 1.5$ Å) and oxygen ($r = 1.4$ Å). The interactions of the other donor atoms with the complexed cation are therefore weakened and the difference between calculated and reaction enthalpies measured increases.

8.4   LEAD ION COMPLEXES

The Pb^{2+} cation forms very stable complexes with crown ethers, see Table 6.1.1. Since the 'individual bond strength' with a nitrogen donor atom is stronger than for an oxygen atom, see Table 5.2.1, one would expect even stronger complexes with diaza crown ethers. At first glace the results summarized in Table 8.4.1 confirm this prediction.

A more careful inspection shows some important differences in the results for the ligand 18C6. The value of the enthalpy in the case of 18C6 is higher than that for the analogue diaza crown ether (22). This reduction is attributed, as already mentioned in detail, to structural changes in the diaza ligand. As a result of favourable entropic contributions the complex formed with (22) is stronger

TABLE 8.4.1

Stability constants and thermodynamic parameters for the reaction of Pb^{2+} with diaza crown ethers and cryptands in methanol at 25 °C (from ref. 71, 133 and 147)

Ligand	Log K[a]	$-\Delta H$[b]	$T\Delta S$[b]
DAOO[c]	6.03	33.8	0.5
(21)	6.71	18.1	20.0
(22)	9.11	29.1	22.7
(22DD)	8.37	73.5	-25.9
(23)	7.94	33.2	11.9
(211)	9.03	24.6	26.7
(221)	12.84	67.9	5.1
(222)	12.95	72.7	0.9
(222B)	12.22	61.2	8.2
(222BB)	10.90	52.7	9.2
(211DL)	-		
(221DL)	3.57	5.2	15.1
(222DL)	5.39	17.9	12.7

[a] K in M^{-1}
[b] ΔH and $T\Delta S$ in kJM^{-1}
[c] $H_2N-CH_2-CH_2-O-CH_2-CH_2-O-CH_2-CH_2-NH_2$ (DAOO), from ref. 98

than with 18C6. Differences in the solvation of both ligands may be the reason because other factors influencing the measured reaction entropies remain constant. For the reaction of the ligand (22DD) with Pb^{2+} a great increase in the value of the reaction enthalpy accompanied by a decrease in the reaction entropy in comparison with (22) is measured. Identical observations have already been discussed for other cations.

In contrast with the results obtained for Ag^+, increasing stability constants are found going from noncyclic diamines to macrocyclic diaza crown ethers and the macrobicyclic cryptands. The formation of diaza crown ether complexes is even favoured by entropic factors compared with the noncyclic diamine. Enthalpic contributions further increase the stability constants for the cryptands if the cavity is large enough to accommodate the Pb^{2+} cation. Further discussion is not possible due to the uncertainties for the 'individual bond strength' with the oxygen donor atom.

The values of the reaction enthalpies decrease going from the cryptand (222) to (222B) and further to (222BB). This behaviour indicates the existence of a second solvation shell. For other divalent cations identical observations are made.

The complex stabilities of the macrobicyclic dialactam ligands, see Figure 8.1.7, are reduced by a factor of 10^8 or 10^9 in comparison with the corresponding cryptands. Also for this cation the reduction of the stability constants is caused by enthalpic factors only since the reactions are favoured by the reaction entropies.

8.5   COPPER, COBALT AND NICKEL ION COMPLEXES

Experimental results including thermodynamic data for the reactions of transition metal ions with diaza crown ethers and cryptands in methanol are extremely scarce. The parameters for the complexation of Cu^{2+}, Co^{2+} and Ni^{2+} are summarized in Table 8.5.1.

Compared with the noncyclic ligand DAOO the stability of the complexes formed with macrocyclic and macrobicyclic ligands is increased by several orders of magnitude. The few known thermodynamic data indicate that the macrocyclic effect in the case of Cu^{2+} is caused by entropic factors. No statement about the complexes with cryptands is possible at this time.

The complex formation with Co^{2+} and Ni^{2+} with diaza crown ethers and cryptands is favoured by entropic factors. With those ligands positive values for the reaction enthalpies are observed. Thus the formation of complexes is only possible if the contribution of the reaction enthalpy is compensated by the reaction entropy. These complexed cations only interact with the nitrogen donor atoms of the ligands. Structural changes of the ligand during the reaction are responsible for the positive values of the reaction enthalpies. Changes in the solvation of the ligand and the cation cause the liberation of solvent molecules. As a result the values of the reaction entropies are obtained.

The substitution of long alkyl chains (22DD) for both protons of the amino groups of (22) leads to an increase in the reaction enthalpies. This is in accordance with the results obtained for other cations.

Direct experimental evidence for structural changes in the ligand are obtained from the reaction of Ni^{2+} with different cryptands. A two-step mechanism for the complexation reaction is observed (ref. 78). Such a mechanism is supported by NMR (ref. 127),

TABLE 8.5.1

Stability constants and thermodynamic values for the complexation of Cu^{2+}, Co^{2+} and Ni^{2+} by diaza crown ethers and cryptands in methanol at 25 °C.

Ligand	Value[a]	Cu^{2+}	Co^{2+}	Ni^{2+}
DAOO[b]	log K	2.86[c]	2.51[d]	5.88[d]
	$-\Delta H$	55.5[c]	7.2[d]	
	$T\Delta S$	-39.2[c]	7.1[d]	
(21)	log K	9.45[e]	6.9[d]	4.90[d]
	$-\Delta H$	32.6[c]	-5.2[d]	-23.8[d]
	$T\Delta S$	21.1	44.4[d]	51.6[d]
(22)	log K	8.48[e]	3.56[d]	3.90[d]
	$-\Delta H$		-11.4[d]	-24.7[d]
	$T\Delta S$		31.6[d]	46.9[d]
(22DD)	log K		2.36[d]	2.54[d]
	$-\Delta H$		-2.8[d]	-11.2[d]
	$T\Delta S$		16.2[d]	25.6[d]
(23)	log K	>5[c]	3.59[d]	4.04[d]
	$-\Delta H$	65.8[c]	-8.4[d]	-16.5[d]
	$T\Delta S$		28.8[d]	39.5[d]
(211)	log K	9.51[e]	6.38[d]	9.3[d]
	$-\Delta H$		-	-11.6[d]
	$T\Delta S$			64.4[d]
(221)	log K	10.08[e]	13.4[d]	9.6[d]
	$-\Delta H$		-	-11.2[d]
	$T\Delta S$			65.7[d]
(222)	log K	8.59[e]	2.47[d]	6.9[d]
	$-\Delta H$		-8.1[d]	-13.5[d]
	$T\Delta S$		22.1[d]	52.7[d]

[a] K in M^{-1}, ΔH and $T\Delta S$ in kJM^{-1}

[b] $H_2N-CH_2-CH_2-O-CH_2-CH_2-O-CH_2-CH_2-NH_2$, (DAOO)

[c] from ref. 77

[d] from ref. 78

[e] from ref. 148

kinetic (ref. 149) and thermodynamic studies (ref. 135). The reaction can be described by the following steps:

$$M^{2+} + L \; \overset{\Delta H_1}{\rightleftharpoons} \; M^{2+} \cap L \; \overset{\Delta H_2}{\rightleftharpoons} \; M^{2+} \subset L$$

An exclusive complex is denoted by ($M^{2+} \cap L$) and an inclusive complex by ($M^{2+} \subset L$). During the complex formation between Ni^{2+} and a cryptand one exotherm and an endotherm process are observed. No direct association of these results to a special process is possible from the experiment. However, a negative enthalpy is expected in the formation of a Ni^{2+} nitrogen bond. The positive reaction enthalpies should normally account for conformational changes during the formation of inclusive complexes.

Using the 'individual bond strength' from Table 5.2.1 it is possible to obtain the difference between calculated and experimental reaction enthalpies. This is shown in Figure 8.5.1.

The differences found are quite similar to those for other cations. Thus, the same interpretation is obviously valid.

For the reaction of Ni^{2+} with different cryptands it is possible to split ΔH, given in Table 8.5.1, into separate values for ΔH_1 and ΔH_2. These can be compared with the difference Δ between calculated (ΔH_c) and reaction enthalpies measured (ΔH_m). The values are summarized in Table 8.5.2.

TABLE 8.5.2
Reaction enthalpies for the two-step complexation of Ni^{2+} by different cryptands in methanol at 25 $^{\circ}$C and the difference Δ between calculated and measured overall reaction enthalpies.

Ligand	$-\Delta H_1$ [a]	$-\Delta H_2$ [a]	Δ [b]
(211)	22.4	-34.0	-33.6
(221)	26.6	-37.8	-33.2
(222)	25.5	-39.0	-35.5

[a] in kJM^{-1} from ref. 78
[b] in kJM^{-1}

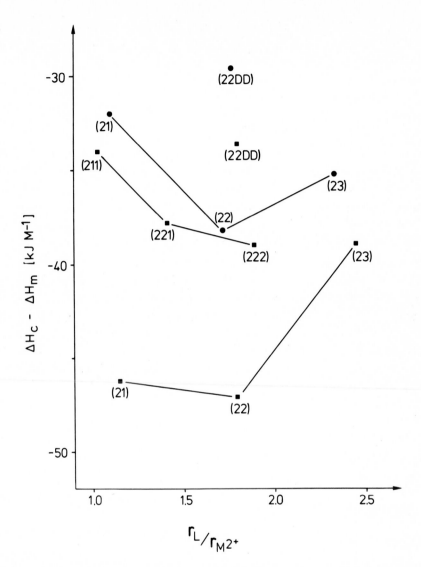

Fig. 8.5.1. Difference between calculated (ΔH_c) and reaction enthalpies measured (ΔH_m) as a function of the ligand to cation ratio (Co^{2+} ●, Ni^{2+} ■).

176

The experimental values of ΔH_2 and the calculated values Δ are well in accordance with one another. Thus, the conformational changes of the ligands during the reactions are the dominant factor in the reaction enthalpies observed. The activation energy of a conformational change for macrobicyclic diamines was found to be 32.2 kJM^{-1} (ref. 15).

9   CONCLUSION

A simple model was used to discuss origin of the macrocyclic and cryptate effects. Many questions cannot be answered at this stage. As an example, differences between the reaction enthalpies of the diaza crown ethers (22DD) and (22) differ from one cation to another. Further research into special problems like the influence of solvent is necessary. With an increasing number of experimental results, e.g. stability constants and thermodynamic data, a more sophisticated interpretation will become possible. It is hoped that the main factors influencing the complex stabilities of crown ethers and cryptands have been discussed.

ACKNOWLEDGEMENTS

The author would like to acknowledge Professor K.H. Drexhage's support throughout this project. He would also like to thank the Deutsche Forschungsgemeinschaft for their financial support and Dr. R. Klink (Merck, Darmstadt) for the donation of non-commercially available ligands. The author is grateful to Ms. K. Baudis for her excellent drawings and to Mrs. R. Hiller for her assistance with the English.

REFERENCES

1 C.J. Pedersen, J. Am. Chem. Soc., 89 (1967), 7017; ibid, 92 (1970), 386.

2 A. Werner, Z. Anorg. Allgem. Chem., 3 (1893), 267.

3 G. Schwarzenbach, Helv. Chim. Acta, 35 (1952), 2344.

4 A.E. Martell, in Advances in Chemistry Series, Vol. 62, R.F. Gould (ed.), (1967), 272.

5 D. Vorländer, Justus Liebigs Ann. Chem., 280 (1894), 167.

6 J.W. Hill and W.H. Carothers, J. Am. Chem. Soc., 57 (1935), 925.

7 E.W. Spanagel and W.H. Carothers, J. Am. Chem. Soc., 57 (1935), 929.

8 D.G. Stewart, D.Y. Waddan and E.T. Borrows, British Patent 785, 229 (23.10.1957).

9 K. Ziegler and R. Aurnhammer, Liebigs Annal. Chem., 513 (1934), 43. A. Lüttringhaus and K. Ziegler, Liebigs Annal. Chem., 528 (1937), 155. K. Ziegler, A. Lüttringhaus and K. Wohlgemuth, 528 (1937), 162. A. Lüttringhaus, Liebigs Annal. Chem., 528 (1937), 181, 211 and 223. A. Lüttringhaus and J. Sichert-Modrow, Makromol. Chem. 18/19 (1967), 511.

10 J.L. Down, J. Lewis, B. Moore and G. Wilkinson, J. Chem. Soc., (1959), 3767.

11 C. Moore and B.C. Pressman, Biochem. Biophys. Res. Commun. 15 (1964), 562. B.C. Pressman, Proc. Natl. Acad. Sci., 53 (1965), 1076. B.C. Pressman, E.J. Harris, W.S. Jagger and J.H. Johnson, Proc. Natl. Sci., 58 (1967), 1949.

12 R.M. Izatt, J.H. Tytting, D.P. Nelson, B.L. Haymore and J.J. Christensen, Science, 164 (1969), 443. R.M. Izatt, J.H. Rytting, D.P. Nelson, B.L. Haymore and J.J. Christensen, J. Am. Chem. Soc. 93 (1971), 1619.

13 H.K. Frensdorff, J. Am. Chem. Soc., 93 (1971), 600.

14 K.H. Wong, K. Konizer and J. Smid, J. Am. Chem. Soc., 92 (1970), 666.

15 C.H. Park and H.E. Simons, J. Am. Chem. Soc., 90 (1968), 2428, 2429 and 2431.

16 B. Dietrich, J.M. Lehn and J.P. Sauvage, Tetrahedron Lett., (1969), 2885 and 2889.

17 J.M. Lehn and J.P. Sauvage, J. Chem. Soc., Chem. Commun., (1971), 440.

18 G.W. Gokel and S.H. Korzeniowski, Macrocyclic Polyether Syntheses, Springer, Berlin 1982.

19 R.M. Izatt, J.S. Bradshaw, S.A. Nielsen, J.D. Lamb and
   J.J. Christensen, Chem. Rev., 85 (1985), 271.

20 D.K. Cabbiness and D.W. Margerum, J. Am. Chem. Soc., 91 (1969),
   6540.

21 F.P. Hinz and D.W. Margerum, J. Am. Chem. Soc., 96 (1974), 4993
   and Inorg. Chem., 13 (1974), 2941.

22 L. Fabbrizzi, P. Paoletti and A.B.P. Lever, Inorg. Chem., 15
   (1976), 1502.

23 M. Kodama and E. Kimura, J. Chem. Soc., Chem. Commun., (1975),
   326 and 891.

24 P. Paoletti, in Bioenergetics and Thermodynamics: Model Systems,
   A. Braibanti (ed.), Reidel, Dortrecht, 1980, 93 pp. and in
   Thermochemistry and its Applications to Chemical and Biochemical
   Systems, M.A.V. Ribeiro da Silva (ed.), Reidel, Dortrecht, 1984,
   339 pp.

25 L.L. Diaddario, L.L. Zimmer, T.E. Jones, L.S.W.L. Sokol, R.B.
   Cruz, E.L. Yee, L.A. Ochrymowycz and D.B. Rorabacher, J. Am.
   Chem. Soc., 101 (1979), 3511. L.S.W.L. Sokol, L.A. Ochrymowycz
   and D.B. Rorabacher, Inorg. Chem., 20 (1981), 3189.

26 M. Kodama and E. Kimura, Bull. Chem. Soc. Jpn., 49 (1976), 2465.

27 B.L. Haymore, J.D. Lamb, R.M. Izatt and J.J. Christensen,
   Inorg. Chem., 21 (1982), 1598.

28 F. Arnaud-Neu, M.-J. Schwing-Weill, R. Louis and R. Weiss,
   Inorg. Chem., 18 (1979), 2956.

29 R.M. Izatt, R.E. Terry, L.D. Hansen, A.G. Avondet, J.S. Bradshaw,
   N.K. Dalley, T.E. Jensen, B.L. Haymore and J.J. Christensen,
   Inorg. Chim. Acta, 30 (1978), 1.

30 G. Anderegg, Helv. Chim. Acta, 58 (1975), 1218.

31 E. Kauffmann, J.-M. Lehn and J.-P. Sauvage, Helv. Chim. Acta,
   59 (1976), 1099.

32 R.M. Izatt, D.J. Eatough and J.J. Christensen, Struct. Bonding,
   16 (1973), 161. J.D. Lamb, R.M. Izatt, J.J. Christensen and
   D.J. Eatough, in Coordination Chemistry of Macrocyclic Compounds,
   G.A. Melson (ed.), Plenum, 1979, 145 pp. F. de Jong and D.N.
   Reinhoudt, Stability and Reactivity of Crown-Ether Complexes,
   Adv. Phys. Org. Chem. Vol. 17, Academic Press, London, 1981.
   A.E. Martell, in Development of Iron Chelators for Clinical Use,
   A.E. Martell, W.F. Anderson and D.G. Badman (eds.), Elsevier,
   Amsterdam, 1981, 67 pp. M. Hiraoka, Crown Compounds their Charac-
   teristics and Applications, Elsevier, Amsterdam, 1982. B. Diet-
   rich, in Inclusion Compounds Vol. 2, J.L. Atwood, J.E.D. Davies

and D.D. MacNicol (eds.), Academic Press, London, 1984, 337 pp.

33 W.E. Morf and W. Simon, Helv. Chim. Acta, 54 (1971), 794.

34 W.E. Morf and W. Simon, Helv. Chim. Acta, 54 (1971), 2683.
    W. Simon and W.E. Morf, in Membranes Vol. 2, G. Eisenman (ed.),
    M. Dekker, New York, 1973, 329 pp. W. Simon, W.E. Morf and
    P.C. Meier, Struct. Bonding, 16 (1973), 113. W.E. Morf,
    D. Ammann, R. Bissig, E. Pretsch and W. Simon, in Progress in
    Macrocyclic Chemistry Vol. 1, R.M. Izatt and J.J. Christensen
    (eds.), J. Wiley, New York, 1979, 1 pp.

35 J.D. Lamb, R.M. Izatt and J.J. Christensen, in Progress in
    Macrocyclic Chemistry Vol. 2, R.M. Izatt and J.J. Christensen
    (eds.), J. Wiley, New York, 1981, 41 pp.

36 A. Pullman, C. Giessner-Prettre and Yu.V. Kruglyak, Chem. Phys.
    Lett., 35 (1975), 156. T. Yamabe, K. Hori, K. Akagi and K. Fukui,
    Tetrahedron, 35 (1979), 1065. V.B. Volkov and K.B. Yatsimirski,
    Teor. Eksp. Khim., 15 (1979), 711.

37 G. Wipff, P. Weiner and P. Kollman, J. Am. Chem. Soc., 104
    (1982), 3249. G. Wipff and P. Kollman, Nouv. J. Chim., 9 (1985),
    457.

38 S.V. Hannongbua and B.M. Rode, Inorg. Chem., 24 (1985), 2577.

39 A. Pullman, in Physical Chemistry of Transmembrane Ions Motions,
    G. Spach (ed.), Elsevier, Amsterdam, 1983, 153 pp. N. Gresh and
    A. Pullman, in Metal Ions in Biological Systems Vol. 19, H. Sigel
    (ed.), M. Dekker, New York, 1985, 335 pp.

40 M. Born, Z. Phys., 1 (1920), 45. T. Abe, J. Phys. Chem., 90
    (1986), 713.

41 see for example: J. Burgess, Metal Ions in Solution, E. Horwood,
    Chichester, 1978. H.L. Friedman and C.V. Krishnan, in Water a
    Comprehensive Treatise Vol. 3, F. Franks (ed.), Plenum, New York,
    1973, 1 pp. P. Schuster, W. Jakubetz and W. Marius, Top. Curr.
    Chem., 60 (1975), 1. B.E. Conway, Ionic Hydration in Chemistry
    and Biophysics, Elsevier, Amsterdam, 1981.

42 F. Vögtle, H. Sieger and W.M. Müller, Top. Curr. Chem., 98
    (1981), 107.

43 G.W. Gokel, D.J. Cram, C.L. Liotta, H.P. Harris and F.L. Cook,
    J. Org. Chem., 39 (1974), 2445.

44 F. Vögtle, W.M. Müller and E. Weber, Chem. Ber., 113 (1980),
    1130.

45 J.A.A. de Boer, D.N. Reinhoudt, S. Harkema, G.J. van Hummel and
    F.J. de Jong, J. Am. Chem. Soc., 104 (1982), 4073. E. Weber,
    S. Franken, H. Puff and J. Ahrendt, J. Chem. Soc., Chem. Commun.,

(1986), 467.

46 P.A. Mosier-Boss and A.I. Popov, J. Am. Chem. Soc., 107 (1985),
6168 and references therein.

47 J. Dzidič and P. Kebarle, J. Phys. Chem., 74 (1970), 1466.

48 A.E. Martell, in Advances in Chemistry Series Vol. 62, R.F. Gould
(ed.), American Chemical Society, Washington, 1967, 272 pp.

49 A.I. Popov and J.M. Lehn, in Coordination Chemistry of Macrocyclic Compounds, G.A. Melson (ed.), Plenum, New York, 1979,
537 pp.

50 J. Gutknecht, H. Schneider and J. Stroka, Inorg. Chem., 17
(1978), 3326.

51 G.W. Gokel, D.M. Goli, C. Minganti and L. Echegoyen, J. Am. Chem.
Soc., 105 (1983), 5786.

52 R.M. Izatt, R.E. Terry, D.P. Nelson, Y. Chan, D.J. Eatough,
J.S. Bradshaw, L.D. Hansen and J.J. Christensen, J. Am. Chem.
Soc., 98 (1976), 7626.

53 J.D. Lamb, R.M. Izatt, C.S. Swain and J.J. Christensen, J. Am.
Chem. Soc., 102 (1980), 476.

54 B.L. Haymore, J.D. Lamb, R.M. Izatt and J.J. Christensen,
Inorg. Chem., 21 (1982), 1598.

55 H.-J. Buschmann, Thermochim. Acta, 102 (1986), 179.

56 Y. Marcus, Ion Solvation, J. Wiley, New York, 1985.

57 H.-J. Buschmann, unpublished results.

58 G. Åkerlöf and O.A. Shorf, J. Am. Chem. Soc., 58 (1936), 1241.
R.G. Bates and R.A. Robinson, in Chemical Physics of Ionic
Solutions, B.E. Conway and R.G. Barradas (eds.), J. Wiley,
New York, 1966, 211 pp.

59 B.G. Cox, C. Guminski and H. Schneider, J. Am. Chem. Soc.,
104 (1982), 3789. B.G. Cox, P. Firman, D. Gudlin and H.Schneider,
J. Phys. Chem., 86 (1982), 4988.

60 B.G. Cox, J. Stroka, P. Firman, I. Schneider and H. Schneider,
Aust. J. Chem. 36 (1983), 2133. B.G. Cox, J. Stroka, P. Firman,
I. Schneider and H. Schneider, Z. Phys. Chem. N.F., 139 (1984),
175.

61 H. Strehlow and H.M. Koepp, Z. Electrochem., 62 (1958), 373.
H. Schneider and H. Strehlow, Z. Phys. Chem. N.F., 49 (1966),
44.

62 M.H. Abraham, A.F. Danil de Namor, W.H. Lee, J. Chem. Soc.,
Chem. Commun., (1977), 893. M.H. Abraham and H.C. Ling, Tetrahedron Lett., 23 (1982), 469.

63 L. Favretto, Ann. Chim., 66 (1976), 621; ibid, 71 (1981), 163.

G. Ercolani, L. Mandolini and B. Masci, J. Am. Chem. Soc., 103 (1981), 7484.

64 G. Chaput, G. Jeminet and J. Juillard, Can. J. Chem., 53 (1975), 2240.

65 P.U. Früh and W. Simon, in Protides of the Biological Fluids - 20th Colloquium, H. Peeters (ed.), Pergamon, Oxford, 1973, 505 pp.

66 H.-J. Buschmann, Z. Phys. Chem. N.F., 139 (1984), 113.

67 H.-J. Buschmann, Inorg. Chim. Acta, 105 (1985), 59.

68 H.-J. Buschmann, Polyhedron, 4 (1985), 2039.

69 H.-J. Buschmann, Makromol. Chem., 187 (1986), 423.

70 K. Ono, H. Konami, K. Murakami, R PPP J, 22 (1979), 19.

71 H.-J. Buschmann, Inorg. Chim. Acta, 98 (1985), 43.

72 H.-J. Buschmann, Inorg. Chim. Acta, 102 (1985), 95.

73 F. Vögtle and E. Weber, Angew. Chem., 91 (1979), 813. B. Tümmler, G. Maass, E. Weber, W. Wehner and F. Vögtle, J. Am. Chem. Soc., 99 (1977), 4683. B. Tümmler, G. Maass, F. Vögtle, H. Sieger, U. Heimann and E. Weber, J. Am. Chem. Soc., 101 (1979), 2588.

74 J. Bjerrum, Metal Amine Formation in Aqueous Solution, Haase, Copenhagen, 1941. P. Paoletti, L. Fabrizzi and R. Barbucci, Inorg. Chim. Acta, Rev., 7 (1973), 43. G.A. Melson, in Coordination Chemistry of Macrocyclic Compounds, G.A. Melson (ed.), Plenum, New York, 1979, 17 pp.

75 R.M. Smith and A.E. Martell, Critical Stability Constants Vol. 1, Plenum, New York, 1975.

76 H.-J. Buschmann, Thermochim. Acta, 107 (1986), 219.

77 H.-J. Buschmann, in preparation.

78 H.-J. Buschmann, Inorg. Chim. Acta, 108 (1985), 241.

79 N.K. Dalley, in Synthetic and Multidentate Macrocyclic Compounds, R.M. Izatt and J.J. Christensen (eds.), Academic Press, New York, 1978, 207 pp.

80 R.D. Shannon and C.T. Prewitt, Acta Cryst., B 25 (1969), 925; ibid, B 26 (1970), 1046.

81 M. Dobler, Chimia, 38 (1984), 415.

82 C.C. Chen and S. Petrucci, J. Phys. Chem., 86 (1982), 2601.

83 M. Dobler, Ionophores and Their Structures, J. Wiley, New York, 1981.

84 H.-J. Buschmann, Chem. Ber., 118 (1985), 2746.

85 M. Shamsipur and A.I. Popov, J. Am. Chem. Soc., 101 (1979), 4051. H.D.H. Stöver, L.J. Maurice, A. Delville and C. Detellier, Polyhedron, 4 (1985), 1091.

86 H.-J. Buschmann, J. Solution Chem., in press.

87 H.-J. Buschmann, in preparation.

88 H.-J. Buschmann, Chem. Ber., 118 (1985), 4297.

89 T. Miyazaki, S. Yanagida, A. Itoh and M. Okahara, Bull. Chem.
    Soc. Jpn., 55 (1982), 2005. Y. Nakatsuji, T. Nakamura, M.Okahara,
    D.M. Dishong and G.W. Gokel, Tetrahedron Lett., 23 (1982), 1351
    and J. Org. Chem., 48 (1983), 1237. R.A. Schultz, D.M. Dishong
    and G.W. Gokel, J. Am. Chem. Soc., 104 (1982), 625. Y. Nakatsuji,
    T. Nakamura and M. Okahara, Chem. Lett., (1982), 1207. D.M.
    Dishong, C.J. Diamond, M.J. Cinoman and G.W. Gokel, J. Am. Chem.
    Soc., 105 (1983), 586. Y. Nakatsuji, T. Mori and M. Okahara,
    Tetrahedron Lett., 25 (1984), 2171. A. Kaifer, D.A. Gustowski,
    L. Echegoyen, V.J. Gatto, R.A. Schultz, T.P. Cleary, C.R. Morgan,
    D.M. Goli, A.M. Rios and G.W. Gokel, J. Am. Chem. Soc., 107
    (1985), 1958.

90 R.B. Davidson, R.M. Izatt, J.J. Christensen, R.A. Schultz,
    D.M. Dishong and G.W. Gokel, J. Org. Chem. 49 (1984), 5080.

91 K. Kimura, T. Maeda and T. Shono, Anal. Lett., A11 (1978), 821.
    K. Kimura, A. Ishikawa, H. Tamura and T.Shono, J. Chem. Soc.,
    Perkin Trans. II, (1984), 447.

92 J. Ikeda, T. Katayama, M. Okahara and T. Shono, Tetrahedron
    Lett., 22 (1981), 3615.

93 T.M. Handyside, J.C. Lockhart, M.B. McDonnell and P.V. Subba Rao,
    J. Chem. Soc., Dalton Trans., (1982), 2331.

94 S. Shinkai and O. Manabe, Top. Curr. Chem., 121 (1984), 67.

95 F. Vögtle and W. Weber, Angew. Chem., Int. Ed. Engl., 18 (1979),
    753. E. Weber and F. Vögtle, Inorg. Chim. Acta, 45 (1980), L65.

96 J. Bouquant, A. Delville, J. Grandjean and P. Laszlo, J. Am.
    Chem. Soc., 104 (1982), 686.

97 J.D. Lin and A.I. Popov, J. Am. Chem. Soc., 103 (1981), 3773.

98 H.-J. Buschmann, unpublished results.

99 J. Rebek, T. Costello, L. Marshall, R. Wattley, R.G. Gadwood
    and K. Onan, J. Am. Chem. Soc., 107 (1985), 7481.

100 A.C. Coxon and J.F. Stoddart, J. Chem. Soc., Perkin Trans. I,
    (1977), 767. D.G. Parsons, J. Chem. Soc., Perkin Trans. I,
    (1978), 451. J.A. Bandy, D.G. Parsons and M.R. Truter, J. Chem.
    Soc., Chem. Commun., (1981), 729. D.G. Parsons, J. Chem. Soc.,
    Perkin Trans. I, (1984), 1193.

101 R.M. Izatt, G.A. Clark, J.D. Lamb, J.E. King and J.J. Christensen,
    Thermochim. Acta, 97 (1986), 115.

102 G.E. Maas, J.S. Bradshaw, R.M. Izatt and J.J. Christensen,

J. Org. Chem., 42 (1977), 3937. J.S. Bradshaw, G.E. Maas,
R.M. Izatt and J.J. Christensen, Chem. Rev., 79 (1979), 37.
J.S. Bradshaw, R.E. Asay, S.L. Baxter, P.E. Fore, S.T. Jolley,
J.D. Lamb, G.E. Maas, M.D. Thompson, R.M. Izatt and J.J.
Christensen, Ind. Eng. Chem. Prod. Res. Dev., 19 (1980), 86.

103 K. Matsushima, N. Kawamura, Y. Nakafsuji and M. Okahara,
Bull. Chem. Soc. Jpn., 55 (1982), 2181.

104 J.D. Lamb, R.M. Izatt, C.S. Swain, J.S. Bradshaw and J.J.
Christensen, J. Am. Chem. Soc. 102 (1980), 479.

105 G.R. Newkome, J.D. Sauer, J.M. Roper and D.C. Hager, Chem. Rev.,
77 (1977), 513.

106 J.S. Bradshaw, S.L. Baxter, J.D. Lamb, R.M. Izatt and J.J.
Christensen, J. Am. Chem. Soc., 103 (1981), 1821.

107 R.M. Izatt, J.D. Lamb, R.E. Asay, G.E. Maas, J.S. Bradshaw and
J.J. Christensen, J. Am. Chem. Soc., 99 (1977), 6134.

108 H.-J. Buschmann, Z. anorg. allg. Chem., 523 (1985), 107.

109 R.M. Izatt, R.E. Terry, L.D. Hansen, A.G. Avondet, J.S. Bradshaw
N.K. Dalley, T.E. Jensen and B.L. Haymore, Inorg. Chim. Acta,
30 (1978), 1.

110 J.D. Lamb, R.M. Izatt, S.W. Swain and J.J. Christensen, J. Am.
Chem. Soc., 102 (1980), 475.

111 M.L. Campbell, N.K. Dalley, R.M. Izatt and J.D. Lamb, Acta
Cryst., B37 (1981), 1664.

112 B. Dietrich, J.-M. Lehn, J.P. Sauvage and J. Blanzat, Tetrahedron, 29 (1973), 1629.

113 H. Schneider, S. Rau and S. Petrucci, J. Phys. Chem., 85
(1981), 2287.

114 D. Moras and R. Weiss, Acta Cryst., B29 (1973), 396. B. Metz,
D. Moras and R. Weiss, Acta Cryst., B29 (1973), 1377. B. Metz,
D. Moras and R. Weiss, Acta Cryst., B29 (1973), 1382.

115 J.-M. Lehn, J.P. Sauvage and B. Dietrich, J. Am. Chem. Soc.,
92 (1970), 2916.

116 B.G. Cox, D. Knop and H. Schneider, J. Am. Chem. Soc., 100
(1978), 6002. R. Pizer, J. Am. Chem. Soc., 100 (1978), 4239.

117 B.G. Cox and H. Schneider, J. Am. Chem. Soc., 99 (1977), 2809.
B.G. Cox, H. Schneider and J. Stroka, J. Am. Chem. Soc., 100
(1978), 4746.

118 J.-M. Lehn, Struct. Bonding (Berlin), 16 (1973), 1.

119 A. Masuyama, Y. Nakatsuji, J. Ikeda and M. Okahara, Tetrahedron
Lett., 22 (1981), 4665.

184

120 B.D. White, D.D. Dishong, C. Minganti, K.A. Arnold, D.M. Goli and G.W. Gokel, Tetrahedron Lett., 26 (1985), 151.

121 H.-J. Buschmann, Inorg. Chim. Acta, 125 (1986), 31.

122 G.W. Gokel, D.M. Dishong, C.J. Diamond, J. Chem. Soc., Chem. Commun., 22 (1980), 1053; J. Am. Chem. Soc., 105 (1983), 586.

123 R.A. Schultz, D.M. Dishong and G.W. Gokel, J. Am. Chem. Soc., 104 (1982), 625.

124 R.A. Schultz, B.D. White, D.M. Dishong, K.A. Arnold and G.W. Gokel, J. Am. Chem. Soc., 107 (1985), 6559.

125 R.B. Davidson, R.M. Izatt, J.J. Christensen, R.A. Schultz, D.M. Dishong and G.W. Gokel, J. Org. Chem., 49 (1984), 5080.

126 J.-M. Lehn and J.P. Sauvage, J. Am. Chem. Soc., 97 (1975), 6700.

127 E. Mei, A.I. Popov and J.L. Dye, J. Am. Chem. Soc., 99 (1977), 6532. E. Mei, L. Liu, J.L. Dye and A.I. Popov, J. Solution Chem. 6 (1977), 771. E. Kauffmann, J.L. Dye, J.-M. Lehn and A.I. Popov, J. Am. Chem. Soc., 102 (1980), 2274.

128 F. Mathieu, B. Metz, D. Moras and R. Weiss, J. Am. Chem. Soc., 100 (1978), 4912.

129 B.G. Cox, D. Knop and H. Schneider, J. Phys. Chem., 84 (1980), 320.

130 B.G. Cox, P. Firman, I. Schneider and H. Schneider, Inorg. Chim. Acta, 49 (1981), 153.

131 R.M. Lawrence and R.F. Kruh, J. Chem. Phys., 47 (1967), 4758.

132 Y.M. Cahen, J.L. Dye and A.I. Popov, Inorg. Nucl. Chem. Lett., 10 (1974), 899.

133 H.-J. Buschmann, Inorg. Chim. Acta, 120 (1986), 125.

134 A. Hourdakis and A.I. Popov, J. Solution Chem., 6 (1977), 299.

135 H.-J. Buschmann, J. Solution Chem., 15 (1986), 453.

136 B.G. Cox, P. Firman and H. Schneider, Inorg. Chim. Acta, 69 (1983), 161.

137 B.G. Cox, Ng. Van Truong, J. Garcia-Rosas and H. Schneider, J. Phys. Chem. 88 (1984), 996.

138 M.K. Chantooni and J.M. Kolthoff, Proc. Natl. Acad. Sci. U.S.A., 78 (1981), 7245.

139 B.G. Cox, G.R. Hedwig, A.J. Parker and D.W. Watts, Aust. J. Chem. 27 (1974), 477.

140 J.N. Albright, J. Chem. Phys., 56 (1972), 3783.

141 N. Morel-Desrosiers and J.-P. Morel, J. Am. Chem. Soc., 103 (1981), 4743.

142 B. Metz, D. Moras and R. Weiss, J. Am. Chem. Soc., 93 (1971), 1806.

143 B.G. Cox, Ng. Van Truong, J. Garcia-Rosas and H. Schneider, J. Phys. Chem. 88 (1984), 996.

144 H.-J. Buschmann, in preparation.

145 T. Yamaguchi, O. Lindqvist, J.B. Boye and T. Claeson, Acta Chim. Scand., A38 (1984), 423.

146 T. Burchard, Ph. D. Dissertation, Universität Dortmund (F.R.G.), June 1984.

147 H.-J. Buschmann, Chem. Ber., 118 (1985), 3408.

148 B. Spiess, F. Arnaud-Neu and M.-J. Schwing-Weill, Helv. Chim. Acta, 62 (1979), 1531.

149 B. Tümmler, G. Maas, E. Weber, W. Wehner and F. Vögtle, J. Am. Chem. Soc., 99 (1977), 4683. G.W. Liesegang, J. Am. Chem. Soc., 103 (1981), 953. B.G. Cox, J. Garcia-Rosas and H. Schneider, Nouv. J. Chim., 6 (1982), 397.

Chapter 3

STEREOCHEMICAL ASPECTS OF MACROCYCLIC COMPLEXES OF TRANSITION METAL IONS

KAREN E. MATTHES and DAVID PARKER

Department of Chemistry, University of Durham, South Road,
Durham, DH1 3LE (United Kingdom)

1. INTRODUCTION

Over the past fifteen years, a wealth of information has been published
concerning the stereochemistry of transition metal macrocyclic complexes.
This article is concerned primarily with second and third row transition
metal complexes, highlighting the structural aspects of the metal coordination
geometry and of the macrocyclic conformation. The macrocyclic coordination
chemistry of iron and cobalt is now a vast subject, and may be traced through
some leading references [1-3]. Similarly there are innumerable macrocyclic
complexes of copper(II) and nickel(II) [4-6], so the discussion herein is
restricted to some well-defined complexes in unusual oxidation states.

The discussion focusses on those systems in which the metal is bound
within the macrocyclic cavity, usually to at least three of the donors of the
ligand. In doing so, no mention is therefore made to the large class of
supermolecular complexes, in which a transition metal complex is encapsulated -
at least in part - by a coronand, (usually a crown ether). This subject
itself has been recently reviewed, and the term 'second-sphere coordination'
is generally used to define these systems, [7]. Increasing attention is being
paid to the precise conformation that the macrocycle adopts in its metal
complexes and a fine discussion of this subject has been given [8]. The
coverage of this aspect in this article is not intended to be comprehensive and
is limited to selected examples which show unusual features. The literature
has been reviewed up to the middle of 1986 with the aid of the Cambridge
Crystallographic Data File.

2. COPPER AND NICKEL

Copper forms a very large number of macrocyclic complexes and it sometimes
seems that as soon as a new macrocycle is synthesised, attempts are made to
structurally characterise its copper(II) complex. The complexes are usually,
mononuclear, but many dinuclear complexes have been examined [4,9,10]
principally to act as models for some of the blue copper proteins, and as a
model for the enzyme superoxide dismutase, [11]. In the copper proteins

plastocyanin and azurin, for example, the copper ion located at the active
site is reported to have a distorted tetrahedral geometry with two nitrogen
and two sulphur donors, [12]. Synthetic macrocyclic ligands have therefore
been examined in which the ring size, nature of the donor atoms and the
stereochemical constraints imposed upon the cycle by the metal centre,
determine the rate of electron transfer within the complex, [13].

In general, copper(II) prefers a tetragonal coordination geometry while
copper(I) tends to form tetrahedral complexes, [16]. Nickel(II) also forms
a large number of macrocyclic complexes with the most common coordination
geometries being octahedral [5,14], and square planar, [15]. More recently
there has been considerable interest in the ability of macrocyclic ligands to
stabilise the unusual oxidation state nickel(I), and the less common higher
oxidation states, copper(III) and nickel(III).

The examples of nickel(I) complexes have been mainly restricted to N_4
macrocyclic compounds, and the complexes were highly reducing [17,18] unless
the ligand was C or N methylated [19], such as the decamethyl 'cyclam'
derivative, 1. The nickel(I) state will be preferred by ligands which can

R = Me

1   2

selectively define a tetrahedral binding site. Such an arrangement strongly
disfavours the nickel(II) state since the stability of such complexes is very
low. A striking example of a ligand which fulfils these requirements has been
reported, [20]. The catenand, 2 consists of two interlocked coordinating rings,
in which each cycle contains a 2,9-diarylsubstituted phenanthroline subunit.
This macrocycle is ideally suited to tetrahedral geometries and has been shown
to form highly stable copper(I) complexes, [21]. The nickel(I) complex of this
ligand, 3, can be easily made by reduction of the nickel(II) precursor using
weak reducing agents such as an alcohol or ascorbic acid. The nickel(I)
complex is remarkably inert to reoxidation and reacts 10^5 times more slowly
with oxygen than its acyclic analogue, 4. Such a difference may be ascribed
to the unique topology of the two rings - i.e. the interlocking of the two
rings.

$$\underline{3} \qquad \underline{4} \qquad \underline{5}$$

Some interesting results have also been obtained with the nickel complexes of the unsaturated ligand, $\underline{5}$. Reduction of the nickel(II) complex of $\underline{5}$ leads to the initial formation of a metal stabilised ligand radical species. In the presence of an axial π-acceptor ligand ($P(OMe)_3$, CO), the nickel(I) ligand radical species $[Ni(\dot{\underline{5}})]^{o}$ was preferentially stabilised over the nickel(II) species, $[Ni(\dot{\bar{\underline{5}}})]^{+}$, [22].

The replacement of amine ligands with amide groups in macrocyclic poly-amines has yielded a new series of oxo-polyamine ligand, [15]. The macro-cycles - such as $\underline{6}$ and $\underline{7}$ - afford high ligand field strengths to those cations which have ionic diameters compatible with the macrocyclic cavity size. The resultant complexes exhibit high kinetic and thermodynamic stability although some are quite labile in acidic media. The rigid co-planarity and strong in-plane ligand fields of the 12 to 15-membered deprotonated oxo-amine cycles is clearly manifested in the complexation of only the square-planar, low-spin, d^8 Ni^{2+} form. The homologous 12 to 15-membered saturated N_4 cycles yield both low-spin and octahedral, solvated, high-spin Ni^{2+} complexes. It has been demonstrated that the uncommon oxidation states Cu^{3+} and Ni^{3+} [23] may be stabilised by coordination to some of these macrocyclic oxo-polyamines, for example ($\underline{6}$). On the basis of their ESR spectra, the copper(III) complexes were diamagnetic, and the nickel(III)

$$\underline{6} \qquad \underline{7}$$

complexes, low-spin d^7 ($S = \frac{1}{2}$). The overall oxidation potential of macrocyclic M^{2+} complexes affords a measure of the thermodynamic stabilisation of M^{3+}

190

relative to M^{2+} and is a complex function of the macrocyclic cavity size, the number and nature of donor atoms and the type of interacting axial substituents. On going from the smaller dioxo-(13)-ane N_4 to the larger dioxo-(16)-ane N_4 cycle, the oxidation of Ni^{2+} to Ni^{3+} occurs with greater ease whereas the opposite is true for the Cu^{2+}/Cu^{3+} couple. The small sized 13-membered cavity best fits the small Cu^{3+} ion, while the low-spin Ni^{3+} state is stabilised by additional axial ligation and this is most favourable with a larger macrocycle due to its weaker in-plane ligand field. Such an effect is clearly demonstrated with the macrocycle 7 which may accommodate a square-planar N_5 coordination geometry for Ni^{3+}. Indeed it has been suggested that the dioxo-(16)-ane N_5 cycle, 7, represents the most ideal structure for stabilisation of Ni^{3+}. The oxidation potential for the transformation from a high spin, square pyramidal Ni^{2+} to the low-spin, square pyramidal Ni^{3+} configuration is only +0.24V. An X-ray structure of the Ni^{2+} complex of 7 has given further information, [24]. The nickel(II) and four nitrogen donors are nearly perpendicular to the remaining axial nitrogen to

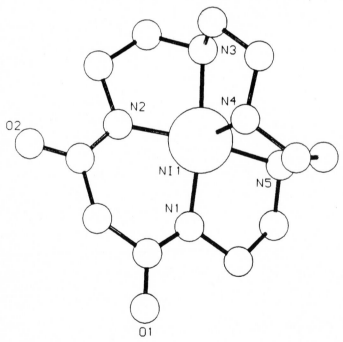

Figure 1. X-Ray crystal structure of the Ni(II) complex of 7[dioxo-(16)-ane N_5]

give a square-pyramidal structure (Figure 1). The ligand is appreciably strained as seen in the narrow chelate bite of the 5-membered rings (NN̂iN averages 83°). It is evident that oxidation or oxygenation of the nickel(II),

concomitant with its shrinkage in size, would relieve such strain, and it has
already been found from the reported X-ray structures of the octahedral Ni^{2+}
and Ni^{3+} complexes of cyclam [25,26], that $N-Ni^{2+}$ bond lengths are longer
$(2.06A^{\circ})$ than $N-Ni^{3+}$ bonds $(1.97A^{\circ})$. This system has attracted much interest
because the nickel(II) complex of 7 binds O_2 in aqueous solution and the
oxygenated complex attacks benzene to yield phenol. The complex is therefore
a rather unique mono-oxygenase model.

3. SILVER AND GOLD

The silver(I) and gold(I) cations are large, polarisable d^{10} cations with
ionic radii of about 1.0 and $1.37A^{\circ}$ respectively in 4-coordinate structures,
[27]. While silver(I) favours a tetrahedral geometry, gold(I) tends to form
stable linear complexes and this feature may be associated with the lack of
gold(I) macrocyclic complexes. Silver(II) is an unusual oxidation state,
which tends to form square planar complexes, while gold(III) forms well
defined square planar d^8 complexes. Generally, silver(I) macrocyclic complexes
are quite common, but there are very few well-defined complexes of Au(I) or
Au(III) with either saturated or unsaturated macrocyclic ligands.

The X-ray structures of the silver(I) complexes of the constitutionally
isomeric, (15)-ring macrocycles 8 and 9 have been reported [28,29].
Complexes were isolated as their thiocyanate salts. In both cases the thio-

8   9

cyanate counterion is directly bound to silver through sulphur. The
coordination geometry about silver may be regarded as a distorted square
pyramid in which the thiocyanate ligand occupies the apical site (Figures 2
and 3). In both cases the metal atom is located in the cavity of the macrocycle and is bound to both of the nitrogen and sulphur atoms in it. With 8,
there is a weak silver-oxygen interaction $(Ag-O, 2.9A^{\circ})$ which is absent with
the isomeric complex of 9.

The silver complex of the $(16)-N_2S_2$ cycle, 10, has been described, [30].
Once again the silver atom has a distorted square-pyramidal coordination,
with axial N atoms and equatorial acetato oxygen and sulphur atoms. In the

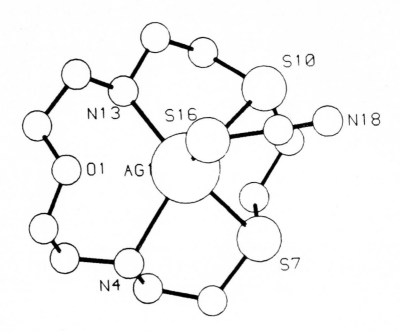

Figure 2. X-Ray crystal structure of the Ag(I) (SCN) complex of 8 [(15)-NS$_2$NO].

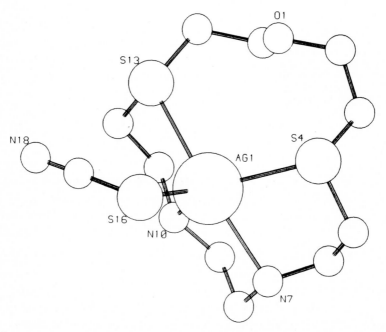

Figure 3. X-Ray crystal structure of the Ag(I) (SCN) complex of 9 [(15)-N$_2$SOS].

complex, all four of the six-membered rings (AgSCCCN) have twist-boat
conformations and the NH groups adopt a cis configuration, (Figure 4).   The

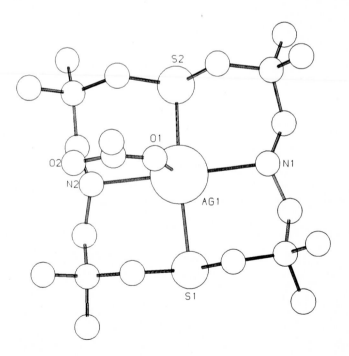

10   11

preference of silver for binding sulphur over oxygen is again illustrated in
the complex with the (18)-ring cycle, 11, [31]. The crystal structure reveals
that each silver cation forms not only an intramolecular Ag-S bond, but also
an intermolecular interaction with an adjacent ligand. One of the silver-
oxygen bonds is relatively short (2.5A°), while the other four are somewhat

Figure 4.   X-Ray crystal structure of the Ag(I) (OCOMe) complex of 10 [(16)-N$_2$S$_2$].

longer (2.66 to 2.92A°), indicative of fairly weak cation-dipole interactions.
The nitrate anion is remote from the complex and does not bind to the silver

194

directly. A similar irregular six-coordinate complex has been defined with
the diiminopyridine ligand, 12, [32]. The silver coordination sphere is
neither octahedral nor trigonal prismatic. It is apparent that the macrocycle

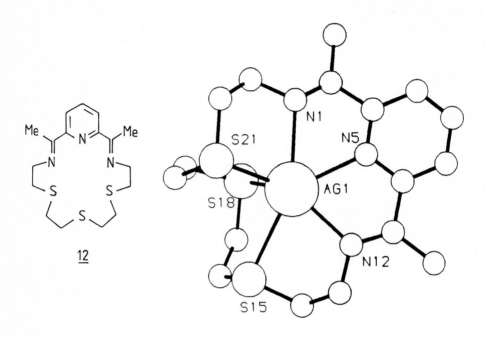

12

Figure 5. X-Ray crystal structure of the Ag(I) complex of 12 [(18)-N_3S_3
diiminopyridine].

adopts a rather strained geometry: one of the Ag-N bonds is 0.20A° longer
than the other, and one of the Ag-S bonds is 0.30A° longer than the other
two (Figure 5). It has been previously noted that in similar complexes with
lead, calcium and strontium [33,34] the macrocycles are close to planarity,
whereas in [12-Ag]$^+$, the central sulphur atom is located 2.32A° above the
AgN$_3$ plane.

The metal assisted cyclisation method has been used widely in the
synthesis of many new macrocyclic ring systems, (e.g. 12). Recently, the
silver catalysed formation of crown ethers has been reported. The reaction of
paraformaldehyde, $(CH_2O)_3$, with AgAsF$_6$ in liquid SO$_2$ gave a complex containing
the [$(CH_2O)_6Ag_2$]$^{2+}$ cation, [35] while a similar reaction with ethylene oxide
gave the sandwich structure illustrated in Figure 6 [36]. Each silver ion
is eight-coordinate and forms weak bonds with each of the eight oxygens
(average Ag-O = 2.57A°). The coordination geometry at silver corresponds to
that of a distorted cube in which two opposite faces have been rotated through
30°. The two (12)-crown-4 ligands adopt conformations similar to other

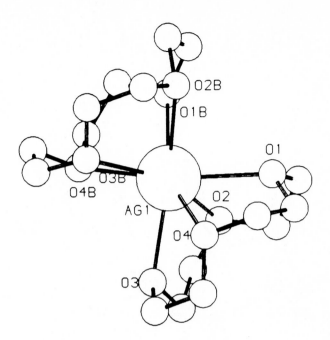

Figure 6. X-Ray crystal structure of the Ag(I) complex of (12)-crown-4.

complexes of this ligand, [8].

An interesting example of the stabilisation of unusual oxidation states by macrocyclic ligands has been found with silver(II). Silver(II) complexes, in general, are rare and are formed with ligands which can tolerate the strong oxidising capability of silver(II). Reaction of silver(I) salts with (14)-ring tetraaza cycles such as 'cyclam' 13 or the hexamethyl derivative 17, leads to disproportionation of the silver(I) to elemental silver and a macrocyclic silver(II) complex [37,38]. Two isomeric complexes were formed with 'cyclam', with the major species having the 'basket' configuration shown in Figure 7a. In this complex the silver ion is displaced 0.24A° below the plane defined by the four nitrogens, and resides in a crystallographic mirror plane bisecting the two six-membered chelate rings. This complex slowly isomerises in solution to the thermodynamically more stable structure shown in Figure 7b. The silver ion sits on a crystallographic centre of symmetry and there are weak axial interactions with perchlorate oxygens (Ag-O 2.79A°) generating a distorted octahedral geometry. A very similar structure was reported for the silver(II) complex of 7, [38].

Macrocyclic complexes of gold are not common, but there is some information on complexes of gold(III) with some unsaturated tetraaza ligands.

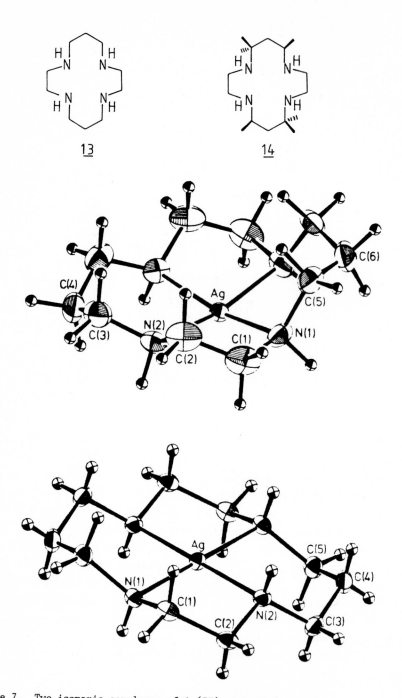

Figure 7.   Two isomeric complexes of Ag(II) with 13 (cyclam).
(a) Main component with basket type configuration.
(b) Minor component, the thermodynamically more stable structure.

Gold(III) ß-diiminate complexes were prepared by reaction of [Au(ethylenediamine)$_2$]Cl$_3$ with a series of ß-diketones in the presence of aqueous base. This reaction generates (14)-ring, tetraaza, 12π macrocyclic complexes, for example 15 and 16 [39,40]. The crystal structures of 15 and 16 are very similar, and in the chloride salt of 15, the cation is very nearly planar with the gold atom at a crystallographic centre of symmetry (Figure 8). The π electrons within the six membered ß-diiminate rings are delocalised as deduced by the observed pattern of C-C and C-N bond distances.

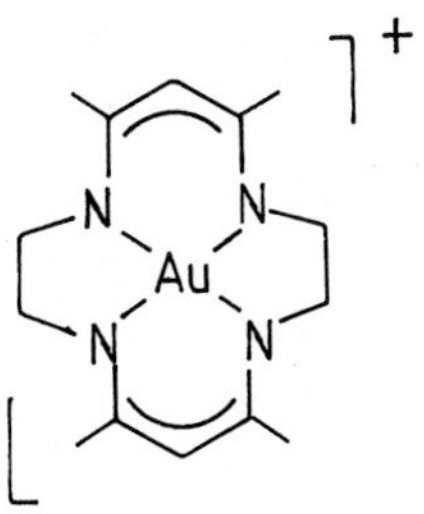

15 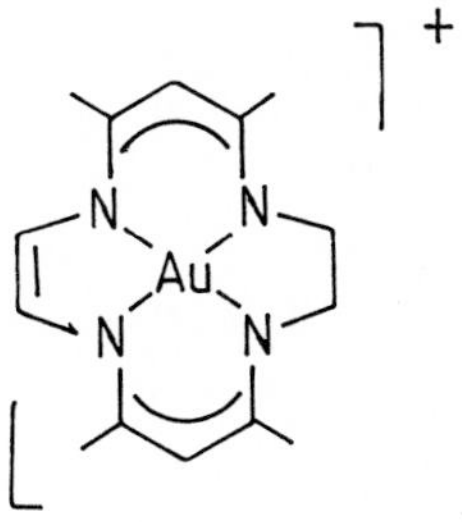   16

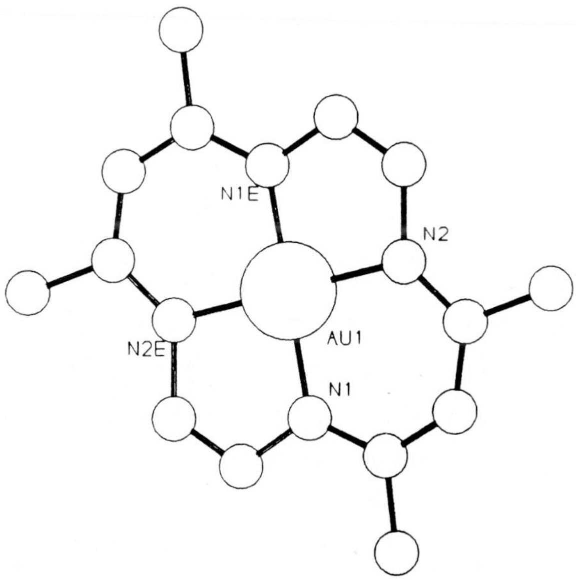

Figure 8.   X-Ray crystal structure of the chloride salt of the Au(III) complex, 15.

198

4. PALLADIUM AND PLATINUM

These two elements are complexed most commonly in the +2 oxidation state, forming square planar d^8 complexes which may be distorted by axial interactions with additional donors to give an approximately square pyramidal geometry. The complexes once formed are kinetically fairly inert - particularly with platinum - but they are often slow to form as well. It is probably this sluggish metal complexation step which may be, in part, associated with the relatively small number of platinum and, to a lesser extent palladium macrocyclic complexes.

One of the earliest complexes to be structurally characterised was the palladium(II)nitrate macrocyclic complex with the (15)-ring cycle 17 [41]. The palladium was located in the cavity of the macrocycle and is strongly bound to the two nitrogen and two sulphur atoms in a distorted square planar array, (Figure 9). The palladium lies 0.23A° above the square plane defined by the N_2S_2 donors, and is bound to the oxygen atom of the cycle (Pd-O = 2.78A°) by electrostatic ion-dipole interactions. The coordination geometry about palladium may therefore be considered to be intermediate in nature between a square plane and a square pyramid, and in this idealised latter state there would be considerable strain in the macrocyclic ring.

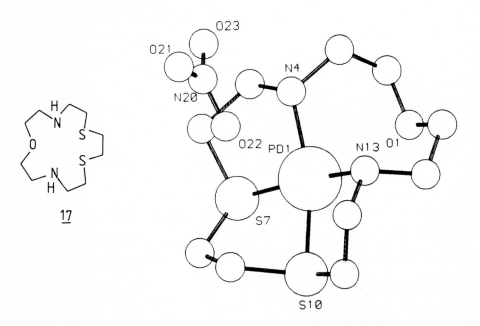

Figure 9. X-Ray crystal structure of the Pd(II) complex of 17 [(15)-NS$_2$NO].

The sixteen-membered macrocycle, 18, forms a set of well-defined palladium(II) complexes, in which once again the palladium atoms have distorted square planar geometry. In two cases a chloride ion resides in fairly distant axial sites. The ligand may adopt one of five different configurations in its square planar complexes, as shown schematically in Figure 10. The positive and negative signs denote the relative orientation of the sulphur lone pairs and NH protons with respect to the ring plane, [42]. The structure of the complex [Pd-(18)]Cl$_2$.2H$_2$O [41] revealed two independent [Pd-(18)]$^{2+}$ cations with quite different configurations. One adopted the ccc configuration with two of the six-membered rings adopting chair conformations and the other two had twist-boat conformations (Figure 11a). In the second cation, only one

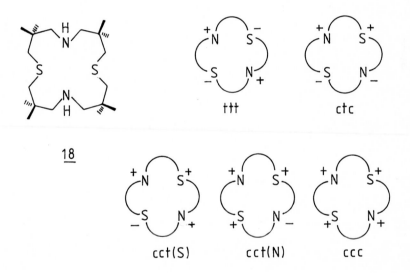

18

Figure 10. The five possible configurations of 18 [(16)-N$_2$S$_2$]; the positive and negative signs denote the relative orientations of the NH protons and the S lone pairs with respect to the ring plane.

ring had a chair conformation and the configuration adopted was cct(N), (Figure 11b). These two configurations may be readily interconverted in solution. These diastereomeric species may be related to the molybdenum complexes of the symmetrical (16)-S$_4$ ligand, discussed in section 8, in which the ligand adopts a ccc configuration with all of the six-membered chelate rings in chair conformations.

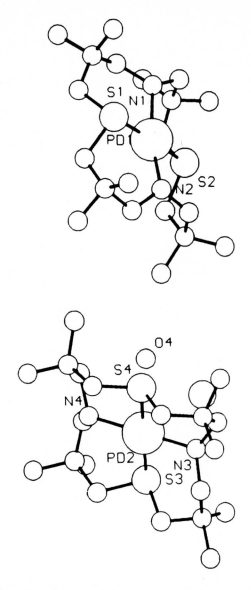

Figure 11. X-Ray crystal structures of two independent Pd(II) complexes of
18 [(16)-N₂S₂].
(a) Cation with ccc configuration.
(b) Cation with cct(N) configuration.

The structure of an unusual one-dimensional palladium(II)-palladium(IV)
mixed valence complex has been reported, involving the ubiquitous-N₄
macrocycle, cyclam, [43]. The palladium(II) complex was prepared by reaction

of the ligand with palladium(II)acetate in water, and the palladium(IV) complex was derived from this by oxidation with nitric acid. The mixed-valence complex was prepared from these two isolated complexes, giving dichroic crystals. The structure of the complex revealed four coordinate palladium(II) and six-coordinate palladium(IV) units stacked alternately to construct a linear chain of Cl-Pd(IV)-Cl---Pd(II)--- segments. The cyclam rings adopted the usual most stable ring conformation, with successive macrocycles rotated 180° about the chain axis.

An unusual example of palladium complexation within a metallomacrocycle has led to the formation of heterotrinuclear chains [44]. The metallomacrocycle, 19, is created using the tridentate ligand bis(diphenylphosphinomethyl)phenylarsine, and the two unbound arsenic atoms and the two metals may bind to a dissimilar metal to create the (14)-ring macrocyclic complex, 20.

$$M = Rh, Ir$$

19   20

In the structure of the dirhodiumpalladium cation, the central palladium ion is bound to both arsenic atoms, but is asymmetrically positioned between the two rhodium ions. One Pd-Rh bond is short (2.70A°), indicative of a single metal-metal bond, whereas the other is much longer (3.17A°) and a bridging chlorine atom links the palladium and rhodium centres, (Figure 12). The two rhodiums are in distinct environments, and the palladium complexation reaction may be regarded as the oxidative addition of a Pd-Cl bond to one of the rhodium centres, giving a six-coordinate geometry about that rhodium.

There are surprisingly few reports of Pt(II) macrocyclic complexes in which all four of the binding sites are occupied by ligand donor atoms. On the other hand, there are several reports of complexes in which three of the donor atoms belong to a macrocyclic ligand. For example diplatinum dichloride complexes 21, 22 and 23 have been characterised [45,46]. However, an example of a rapidly formed Pt(II) complex has been described using the macrocycle 24, which contains two amide ligands. This 'dioxocyclam' ligand encloses a series of metal ions [Cu, Ni, Co and Pd(II)] with simultaneous deprotonation of the two amides to give square planar neutral complexes [47]. A lipophilic

202

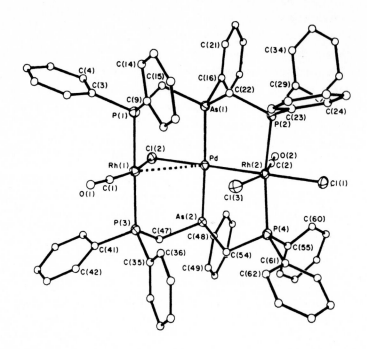

Figure 12.   X-Ray crystal structure of the cation $[Rh_2Pd(CO)_2Cl_3(\mu\text{-}Ph_2PCH_2As\text{-}(Ph)CH_2PPh_2)_2]^+$.

21

22

23

R = H,Me

analogue, 24b, has been proposed as a specific sequestering agent for the extraction of $Pt(NH_3)_2Cl_2$ - the well-known anti-tumour compound.   Finally a brief mention has been made of the square-planar platinum(II) complex 25, prepared after exhaustive reaction of the ligand with $PtCl_2$ in ethanol [48].

a) R = H
b) R = $C_{16}H_{33}$

24

25

5. RHODIUM AND IRIDIUM

Given the importance of organorhodium complexes in homogeneous catalysis, it is rather surprising that there is relatively little work in the literature concerning its macrocyclic complexes. On the other hand, there have been several reports of the second-sphere coordination of rhodium and iridium metal complexes by various crown ethers [7,49-51]. In these complexes, ammonia or acetonitrile ligands bind to the metal centre and interact with an encapsulating crown ether. In a more recent example using an $(18)-N_2O_4$ aza-crown ether, each nitrogen of the (18)-ring cycle is bound to a rhodium(I) cation to give a dirhodium species [52].

Rhodium(I) and iridium(I) favour a square planar coordination geometry, in which they are bound to at least two relatively π-acidic ligands such as a diphosphine or a diene. In the presence of harder donors only, such as nitrogen and oxygen, octahedral rhodium(III) complexes tend to be more stable. Indeed reaction of $RhCl_3$ in methanol with saturated N_4 cycles of varying ring size yields cis or trans octahedral complexes. With (12) and (13)-ring cycles a cis geometry was inferred from electronic spectral data, while with (15) and (16)-ring cycles only a trans geometry was evident. With the (14)-ring, cyclam, a mixture of cis and trans isomers was obtained, [56].

There have been some binding studies in which the complexation of rhodium to phosphine functionalised macrocycles [53] has been examined [34]. A heterodinuclear complex was characterised with 26, in which the zinc ion, bound on top of the $(12)-N_2S_2$ ring, regulated the structure of the ligand by additionally binding to the lateral ether oxygens thereby aiding in the binding of the defined cis-diphosphine unit to a rhodium-norbornadiene cation.

The 2,6 dithiomethylpyridine subunit affords three pre-disposed binding sites for a square planar complex, and rhodium complexes incorporating it have been defined [45]. For example a dicationic dirhodium complex, 27, is formed upon reaction of the parent ligand with $Rh_2Cl_2(CO)_4$ in methanol. A related dirhodium-tricarbonyl complex has been characterised in which the triply bridged $[Rh(CO)_3Rh]^{2+}$ unit is held inside the $(24)-N_6O_2$ macrocycle, (Figure

X = O,S

26

27

13) [55]. The geometry about each Rh(I) ion is distorted octahedral with each
rhodium having the three carbonyl ligands as a common face and additionally
binding to the three nitrogens of each diethylenetriamine subunit on
opposite faces. There is a short rhodium-rhodium single bond (2.58A°). The
complex formed with the hexa-N-methylated analogue of the ligand shown in

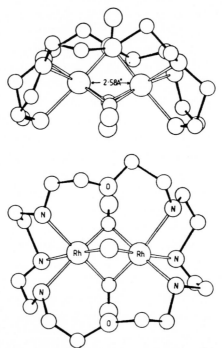

Figure 13. X-Ray crystal structure of the $[Rh(CO)_3Rh]^{2+}$ complex of the (24)-
N_6O_2 macrocycle.

Figure 13, does not have bridging carbonyls. Each rhodium is four coordinate
and sixteen electron. Molecular models indicate that replacing hydrogen by
methyl in the bridged structure would introduce severe steric interactions
between the N(1)-N(3) methyl groups of each N_3 subunit, as well as between
transannular N(2,2') methyl groups.

6. MANGANESE AND CHROMIUM

The complexes of manganese(II) typically show little preference for a particular metal coordination geometry. It is evident, though, that no systematic study of manganese macrocyclic complexes has been undertaken and those complexes which have been reported constitute a fairly diverse set. The high-spin manganese(II) ion is typically associated with a very weak ligand-field stabilisation energy, which probably accounts for the ill-defined coordination geometry. Nonetheless there must be scope for systematically characterising more octahedral Mn(II) complexes in the future. The macrocyclic coordination chemistry of Cr(III) is much more clearly defined and several octahedral complexes have been characterised structurally.

In its complexes with the pentadentate ligands 28 - 30, manganese(II) adopts either distorted pentagonal bipyramid or pentagonal pyramid geometries [57-59]. The donor atoms of the ligand occupy the five equatorial coordination sites. A similar geometry has been found in the seven-coordinate complexes of 29b with iron(II) [60,61]. The unusual heptagonal geometry found may be associated not only with the rigidity of the ligand precursors which - during the Mn(II) templated macrocycle synthesis - do not permit a more conventional coordination geometry but also with the lack of apparent preference for a given coordination geometry of the Mn(II) ion. The structures of the manganese(II) complexes of ligands 28 - 30 are shown in Figures 14 - 16.

28   29   a) n = 3   b) n = 2   30

In the complex with 28, the structure has crystallographic C_2 symmetry with the N_5 equatorial donor set being almost planar (Figure 14). There is a particularly short bond to the coordinated axial perchlorate oxygen. The manganese(II) complex of 29a has been characterised as a 1:1 adduct with 1,10-phenanthroline. The planar phenanthroline molecule is not bound to the metal but is sandwiched between adjacent macrocycle planes in a lamellar structure. It lies above the unsaturated triimine regions of the macrocycle in order to maximise π-π^* interactions. The pentadentate ligand again occupies the pentagonal plane with two water molecules occupying the axial sites. In the macrocycle 30, a 1,10-phenanthroline moiety has been incorporated into the ligand. The

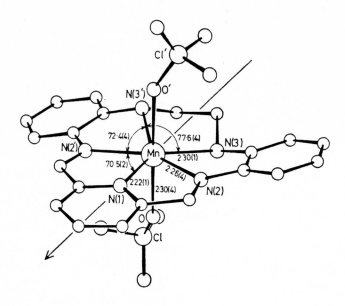

Figure 14.   X-Ray crystal structure of the Mn(II)(ClO$_4$)$_2$ pentagonal bipyramidal complex of <u>28</u> [C$_{21}$H$_{19}$N$_5$] showing the principal bond lengths and angles (e.s.d.s in parentheses).

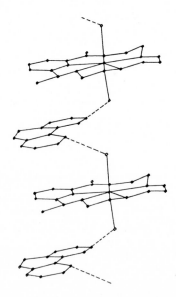

Figure 15.   The lamellar structure of the Mn(II)(H$_2$O)$_2$ complex of <u>29a</u> [(16)-N$_5$ diiminopyridine] and 1,10-phenanthroline. Hydrogen bonds are shown as dotted lines.

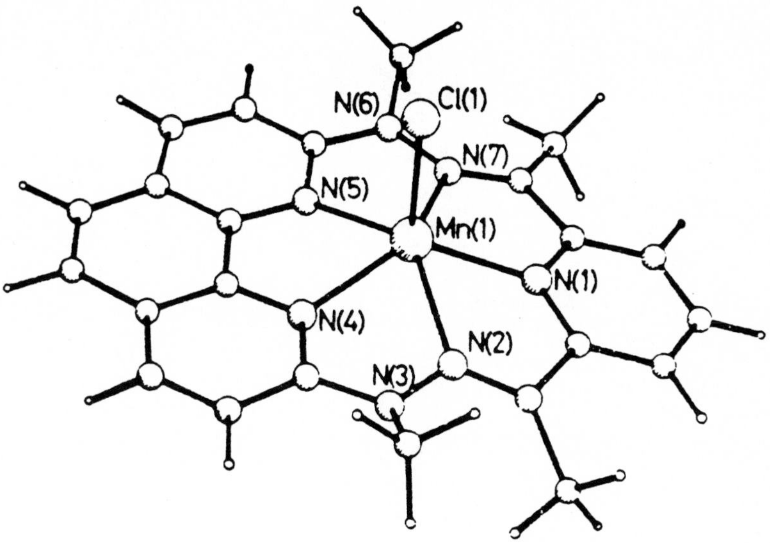

Figure 16.   X-Ray crystal structure of the Mn(II)Cl complex of <u>30</u>
[(15)-N_7 diiminopyridine, 1,10-phenanthroline].

coordination geometry of the Mn(II) complex is a distorted pentagonal
pyramid with the five nitrogen atoms occupying the equatorial base plane
(Figure 16). The manganese(II) ion lies 0.53A$^\circ$ above this plane and a
chlorine atom occupies the axial site. The angles subtended at the metal
centre by adjacent nitrogens vary between 67 and 73° with two Mn-N distances
being appreciably shorter than the others. The rigid phenanthroline moiety
clearly reduces the flexibility of the ligand, although it is still slightly
folded and the pyridine ring makes an angle of only 135° with the plane
through the remaining four nitrogen atoms. The planarity of the ligand and
the shortness of the N-N bonds suggest that π-delocalisation occurs throughout
the entire N_7 ring.

In the complex of manganese with (12)-crown-4, two crown ether ligands
bind in a sandwich manner to give an 8-coordinate structure [62]. The
staggering of the crowns about the metal results in a regular square anti-
prismatic geometry (Figure 17), which may be compared with the more distorted
arrangement in the related silver(I) complex, (section 3). Both of the O_4
rings are coplanar and the Mn-O bond lengths are relatively long indicating
that the coordination is weak, indicative of cation-dipole interactions as
found in many crown ether complexes with alkaline earth metals [8].

The condensation of 2,6-diacetylpyridine with 1,3-diamino-2-hydroxy-

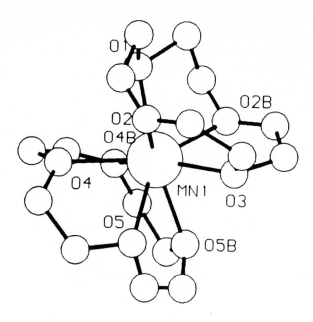

Figure 17. X-Ray crystal structure of the Mn(II) complex of (12)-crown-4.

propane, with a barium(II) template ion, followed by transmetallation with manganese(II) yielded two macrocyclic complexes. One was the dinuclear manganese complex of <u>31</u>, while the minor product was a tetranuclear complex of <u>32</u>, [63]. In the structure of the tetra-nuclear complex each manganese is

<u>31</u>   <u>32</u>

seven-coordinate with an approximate pentagonal bipyramid geometry. In the pentagonal planes, the donors are three nitrogen atoms and two alkoxide oxygen atoms while the axial ligands are a third alkoxide and a semi-coordinated perchlorate anion, (Figure 18). The shortest Mn-Mn distance is 3.32A$^\circ$ precluding any metal-metal bonding. The macrocycle is folded so that the four alkoxide oxygens may cap the faces of the tetrahedral Mn_4 cluster, generating a cubane-like structure. Such Mn_4 clusters are reminiscent of the well studied natural and synthetic Fe_4S_4 clusters.

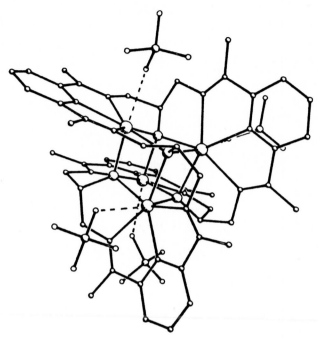

Figure 18. X-Ray crystal structure of the $[Mn(II)(ClO_4)]_4$ complex of 32 [(4x4) Schiff-base].

Chromium(III) macrocyclic complexes have been more systematically studied. The archetypal N_4 macrocycle 'cyclam' (1,4,8,11-tetraazacyclotetradecane) forms a series of cis complexes with a variety of anions, in the monocationic series $[Cr(cyclam)_2X_2]^+$, [64]. More extended investigations with the series of ligands 33 - 40 have permitted the effects of ring size, degree of saturation, number and stereochemistry of alkyl substituents to be examined, [65 - 68]. A Cr(III) complex of 34 with a trans-bis(O-carbamato) structure has been described [65]. The complex was highly symmetric, having a crystallographic centre of symmetry and an approximate mirror plane orthogonal to the CrN_4 plane, containing the symmetry centre and the central carbon of the six-ring, (Figure 19). The five and six membered chelate rings had gauche and chair conformations respectively. Given that the bond lengths and angles between Cr and the nitrogens of the amine were similar to those observed in corresponding ammonia and ethylenediamine complexes, it was suggested that the high spectrochemical Δ value measured (i.e. strong ligand field splitting) may be associated with the rigidity of the ligand rather than

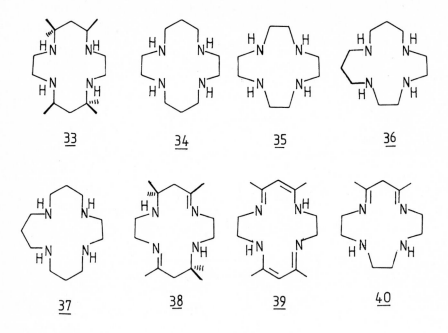

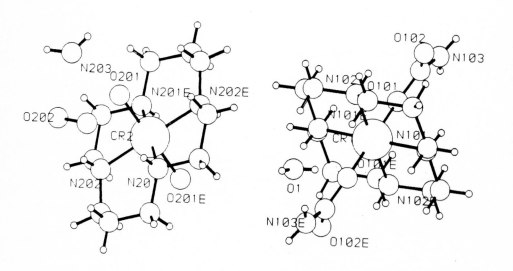

Figure 19. X-Ray crystal structure of the Cr(III) trans-bis(O-carbamato) complex of 34 (cyclam).

to strained Cr-N bonds. Indeed the robust nature of such trans-cyclam species may be related to the inertness of the complex toward axial ligand substitution. Inspection of the structure (Figure 19) suggests that there may well be destabilising steric interactions between an incoming nucleophile and the axial hydrogens of the six-ring chelates. Such interactions would inhibit formation of the transition state structure required for an associative exchange mechanism [69].

The Cr(III) complex with the meso form of 33 has a trans configuration [67], as expected since the ligand could 'fold' only with difficulty due to unfavourable interactions between the substituted methyl groups and the axial ligands [70]. The axial secondary NH protons adopt the meso [RSSR] configuration and the methyl groups are located in pseudo-equatorial positions. The five and six-ring chelates are in gauche and chair conformations and the overall structure is very similar to that shown in Figure 19. In the corresponding dihydroxyl complex of the racemic form of 33, a cis configuration is taken up. The macrocycle folds and the chromium has a distorted octahedral geometry. The two nitrogen atoms adjacent to the dimethyl substituted carbons are trans to each other (Figure 20), [66]. There is a pronounced tetrahedral distortion of the CrN_2O_2 plane owing to interactions of this plane with each equatorial methyl group of the gem-disubstituted carbons.

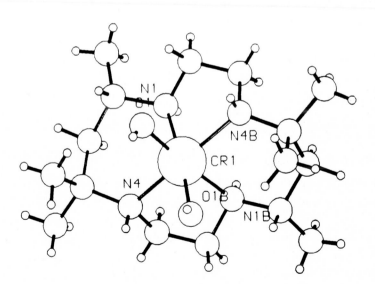

Figure 20. X-Ray crystal structure of the Cr(III) cis-dihydroxyl complex of the racemic form of 33 [(14)-$N_4(Me)_6$].

With the ligands <u>35</u> - <u>37</u>, the cis or trans configuration adopted is a simple function of ring size. With the (12)-ring cycle <u>35</u>, the small cavity enforces a cis configuration, while the larger (15)-ring <u>37</u> only forms a trans complex, [71]. The non-symmetrical (14)-ring ligand <u>36</u> gives equal amounts of cis and trans isomers. Such behaviour may be contrasted with that of <u>34</u>, where the cis complexes are usually isolated in higher yield. It seems likely then that the cis complex of <u>36</u> is less strained than the constitutionally isomeric complexes with the more symmetric ligand <u>34</u>. However the extent of cis/trans isomerism is related to the kinetics of metal binding. The complexes are made under reducing conditions (e.g. zinc in ethanol) and it is a chromium(II) ion which is bound initially. Cr(II) is a larger ion than Cr(III) and initially will bind in a cis manner, thereafter oxidation will be followed by ligand rearrangement to give the thermo-dynamically stable product. Such rearrangements to give trans complexes will obviously occur more easily for larger or more flexible macrocycles - such as a non-symmetric cycle.

With the unsaturated imine cycles, <u>38</u> - <u>40</u>, the N_4 equatorial plane is quite rigidly maintained and trans chromium(III) complexes are uniquely observed [68]. In contrast to the trend observed with a similar series of cobalt(III) complexes, the ligand field strength of the macrocycles bound to Cr(III) decreases as the number of imino nitrogens increases. This may be correlated with the increase in steric strain as the degree of unsaturation rises, associated with the reduction in cavity size.

7. NIOBIUM AND TANTALUM

Only isolated studies of the complexation of salts of these metals with macrocycles have been effected. Reaction of niobium pentachloride with various saturated S_4 and S_6 cycles has been examined and various Lewis acid-base adducts were isolated of varying stoichiometry, [72-74]. In the structure of the adduct with the symmetrical (16)-S_4 cycle, there are two $NbCl_5$ fragments bridged by two sulphur atoms of the macrocycle. The crystal structure was rather disordered and the macrocycle took up an exo configuration, in a similar manner to the free ligand itself, [75]. These studies were rather naively instigated in order to try and characterise niobium macrocyclic complexes: $NbCl_5$ is a hard Lewis acid and the soft sulphur donors of the macrocycles examined do not match. It may be more profitable to examine complexes of niobium in this high oxidation state (d^0) with macrocycles functionalised with ionisable side chains ($-CO_2H$ or $-SH$) in order to reduce the effective nuclear charge at the metal centre.

8. MOLYBDENUM AND TUNGSTEN

Once again only a limited amount of reliable data has been divulged on macrocyclic complexes of molybdenum. Those of tungsten appear non-existent, at this time. The occurrence in nature of sulphur bound molybdenum [76] has promoted the examination of the binding of the lower oxidation state to cyclic thioether ligands. The first complexes to be structurally characterised consisted of molybdenum in the +2 or +4 oxidation state with 41, [77,78]. Reaction of 41 with tetrakis(triflate)dimolybdenum(II) led to the isolation of three separate complexes, in which the macrocycle consistently was binding in an endo-planar fashion. The complexes had the formulae, $[Mo_2^{II}(SH)_2(\underline{41})_2](CF_3SO_3)_2$, $[Mo_2^{IV}O_2(\underline{41}).OEt](CF_3SO_3)_3$ and $[Mo^{IV}O(SH)(\underline{41})]CF_3SO_3$.

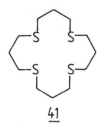

41

The crystal structure of the first of these complexes is shown in Figure 21. The two molybdenum atoms are held in close proximity (Mo-Mo 2.82A°) by the rather unusual sulphur bridges and significant metal-metal interaction must be occurring. Each sulphur in the bridge belongs to an intact cycle. The molybdenum atom is pulled out of the plane of the four sulphurs in the macrocycle to which it is bound, and its position on the macrocycle is off-centre. The Mo-S bond within the bridge is much shorter than the non-bridging Mo-S distances. Each molybdenum is bound to a terminal SH group, which must have arisen during complex formation from some decomposition of the parent cycle, 41. There is some [13]C n.m.r. evidence that this structure is maintained in solution as six resonances were observed in accord with the symmetry deduced from the crystal structure analysis.

In the second complex isolated the cycle coordinates to give a planar S_4 array about the molybdenum. There is a weak, almost linear, oxo-bridge 42 giving rise to the dimeric structure, (Figure 22). The complex is unstable in aqueous solution and electrochemical reduction is irreversible: this instability may be related to the rather weak molybdenum oxygen bond (2.14A°).

The coordination spheres of the two molybdenum atoms differ from each other and from that found in the first complex (Figure 21). The sixth axial site for one molybdenum is an oxo ligand, for the other the site is occupied by a strongly bound ethoxide group. The two macrocyclic rings are pulled away from each other, with one molybdenum lying slightly above and the other

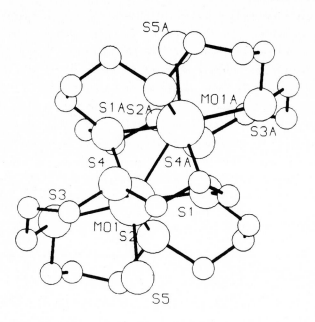

Figure 21. X-Ray crystal structure of the $[Mo(II)]_2(SH)_2$ complex of <u>41</u> $[(16)-S_4]$.

$$\begin{array}{c} \overset{1\cdot76}{} \qquad \overset{1\cdot76A°}{} \\ Et O - Mo = O \cdots\cdots Mo = O \\ \overset{1\cdot86}{} \qquad \overset{2\cdot14A°}{} \end{array}$$

<u>42</u>

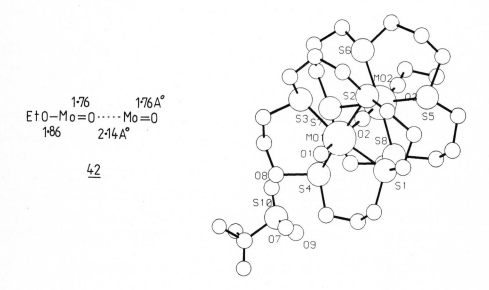

Figure 22. X-Ray crystal structure of the $[Mo(IV)]_2O_2(OEt)$ complex of <u>41</u> $[(16)-S_4]$.

slightly below their respective S_4 planes.

The final complex to be characterised, $[Mo^{IV}O(SH)\underline{41}]CF_3SO_3$, represented the first mononuclear molybdenum macrocyclic complex. The octahedral coordination about the molybdenum differs from that in Figure 21 only in the replacement of a terminal thiol group by an oxo group. The molybdenum lies slightly out of the S_4 plane towards the oxygen atom (Figure 23).

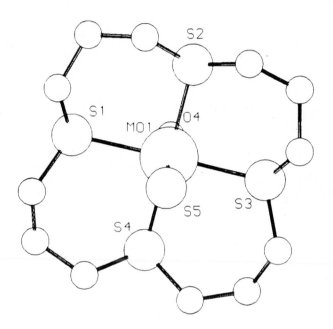

Figure 23. X-Ray crystal structure of the Mo(IV)O(SH) complex of $\underline{41}$ [(16)-S_4].

The reaction of various Mo(III) and Mo(VI) complexes with the symmetric macrocycles (14)-S_4 and (18)-S_6 gave a diverse array of complexes varying from simple monodentate Mo(III) to sulphur coordination to mixed valence complexes involving cleavage of the macrocyclic ring, [79-83]. In all cases the poly-thioethers acted as bridges between molybdenum atoms so that octahedral coordination was maintained, and no evidence was found for a true macrocyclic complex.

9.   TECHNETIUM AND RHENIUM

The γ-emitting radioisotope ^{99m}Tc is widely used in diagnostic nuclear medicine as an imaging agent and in the last few years stable technetium macrocyclic complexes have been developed, [84-86]. The versatility of

216

macrocycles with regard to cavity size, ligand charge and ring substituents
has led to a growing interest in the design of macrocyclic technetium
complexes for specific pharmaceutical use. For example, better brain
scanning agents are being sought in which the technetium complex should be
lipophilic in order to promote passage through the blood-brain barrier. It is
necessary to use the longer-lived ^{99}Tc isotope for structural characterisation
of the complexes, and these studies have been related to those obtained at
tracer levels with the short-lived γ-emitting ^{99m}Tc ($t_{\frac{1}{2}}$ = 6.7 hours).

The most notable complex to be defined is the <u>trans</u>-dioxo(cyclam)
technetium(V) cation which possesses a remarkably symmetrical structure, [87],
(Figure 24). The structure exhibits near-perfect inversion symmetry with the
N_4 rings of the cycle occupying the equatorial plane in the octahedral complex.
The nitrogen configurations are exactly as found in the related complexes
with nickel(III) and nickel(II), [25,26] and copper(II), [88]. The two pairs
of NH protons are oriented axially with respect to the macrocycle, in closer
proximity to the oxygen lone pairs than if they had been equatorially disposed.
The Tc-N bond distances are fairly long to accommodate the fairly small Tc(V)
atom, inducing slight strain into the 'cyclam' ring. The six-ring chelates
are opened up a little and are more puckered chains than in the corresponding

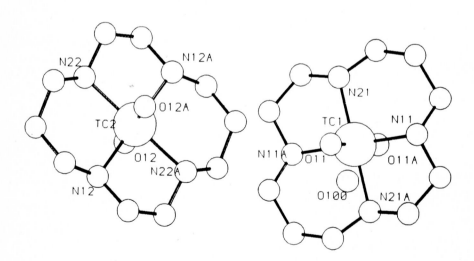

Figure 24.   X-Ray crystal structure of the Tc(V) trans-dioxo complex of <u>34</u>
(cyclam).

complexes of 'cyclam' with first-row transition metals. The high charge
requirements of the Tc(V) centre are satisfied then by the four neutral σ
donor nitrogens and by the two oxo groups. Similar behaviour has been
observed in rhenium chemistry e.g. $[Re(V)O_2(ethylenediamine)_2]^+$ and with the
isoelectronic ruthenium(VI) complexes, (vide infra).

Some acyclic ligands have been described which have oxime functionality
and bind to technetium in a tetradentate manner to give pseudo-macrocyclic
complexes, [89,90]. Some of these complexes are being screened for potential
use as technetium brain scanning agents. With the ligands 43 and 44, the high
charge requirements of the Tc(V) metal centre are satisfied by binding to two
amide nitrogens (i.e. the two amines lose their protons on complexation), one
oximato and one apical oxo group. A strong hydrogen bond between the oxime
oxygens completes the square-pyramidal pseudo-macrocyclic structure. The
apical O=Tc bond is relatively long compared to related monooxo-technetium(V)

43   44

complexes, although still considerably shorter than that found for the trans
dioxo complexes.

There are no authenticated reports of rhenium macrocyclic complexes
although by analogy with the technetium studies already defined, there is
every sign that this will prove a short-lived deficiency.

10. RUTHENIUM AND OSMIUM

While there are now several well-defined examples of octahedral ruthenium
macrocyclic complexes with ruthenium in various oxidation states, the
analogous set of osmium complexes has only just begun to be explored, although
it is likely that further osmium macrocyclic complexes will be reported in the
near future. Complexes of ruthenium(II) and (III) [91,92] with macrocyclic
thioether and secondary amine ligands have been studied, and some saturated
tetradentate tertiary amine cycles have been shown to stabilise the higher
oxidation states of ruthenium [93]. These complexes may find application in
synthetic organic chemistry as catalysts for oxidation reactions.

The structure of the ruthenium(II) complex of 1,4,8,11-tetrathiacyclo-

tetradecane, [(14)-S$_4$], revealed that a cis configuration about the metal
centre was adopted with the ligand binding in an endo cis manner, [91a],
(Figure 25). Thus the previous assignment of configuration made on the basis
of IR evidence [94] was shown to be incorrect, and the structure therefore
resembled related cis complexes of Rh(III) and Co(III), [95]. The
conformation adopted by the cycle in this complex (Figure 25) is very similar
to that observed with the 'cyclam' complex, cis[Co cyclam(ethylenediamine)]$^{3+}$,
[96]. The six-ring chelates adopt chair conformations whilst the 5-rings
take up a distorted gauche form in which the lone pairs on the sulphurs are
mutually trans to each other and are directed toward the cis chlorines. The
Ru-S bonds which were trans to chlorines were 0.07A$^{\circ}$ shorter than those where
sulphur atoms were mutually trans disposed. This expected 'trans effect' may
therefore be a determining factor in overall complex geometry, and it may not
simply be the size of the macrocyclic cavity nor its rigidity which determines
coordination geometry. In the 'folded' form, enhanced π-back bonding between
the metal and the two sulphurs trans to chlorines occurs, leading to
stabilisation of this geometry for the Ru(II) complex.

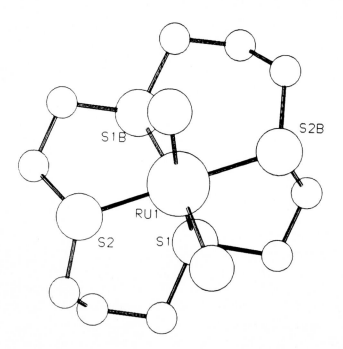

Figure 25.   X-Ray crystal structure of the Ru(II) cis-dichloro complex of
1,4,8,11-tetrathiacyclotetradecane, [(14)-S$_4$].

The reaction of Ru(III) salts with the (14)-N$_4$ cycle, 'cyclam', has been investigated on several occasions, [92a,92c,97-99]. Both cis [92a] and trans [92c] isomers have been structurally characterised, and are shown in <u>Figures 26a and 26b</u>. The trans complex (Figure 26b) has overall C_{2h} symmetry

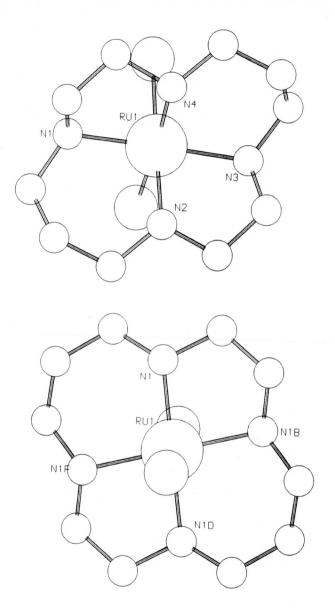

Figure 26. X-Ray crystal structures of both the cis and trans Ru(III) complexes of <u>34</u> (cyclam).
(a) The cis-dichloro cation.
(b) The trans-dichloro cation.

with a ctct (or trans(III)) set of nitrogen configurations, similar to those
of the technetium dioxo complex discussed earlier, (section 9). With both <u>cis</u>
and trans complexes the 6-ring chelates take up chair conformations and the
five rings are all gauche. In the cis isomer, the Ru-Cl bonds are longer
than in the trans, since the chlorines are opposite the nitrogens and
experience a stronger 'trans effect'. Indeed the cis chlorines are more labile
to substitution. In neither structure is a regular octahedral geometry taken
up. In the trans isomer, deviation from planarity occurs in the equatorial
plane, with angles of less than 90° for the five-ring chelates and greater
than 90° (NR̂uN), for the six ring chelates. A further distortion arises from
the fact that the Cl-Ru-Cl axis is not perpendicular to the N_4 plane but tilts
towards the NH atoms, in order to maximise weak intramolecular hydrogen bonds.
In the cis isomer also, the two mutually trans nitrogens lean away from the
axis perpendicular to the Cl-Ru-Cl plane. In this case, weak hydrogen
bonding occurs involving all of the available NH atoms.

A pyridine N_4 cycle, <u>45</u>, has been reported to form a cationic ruthenium-
(II)chlorocarbonyl complex [48] in which the chlorine and carbonyl ligands
are <u>cis</u> disposed and the carbonyl is trans to the pyridine nitrogen, as
observed in several related rhodium-carbonyl complexes with pyridine subunits,
[45,46]. The macrocycle binds in a cis manner and the macrocycle is 'folded'
as previously observed in the complexes of (14)-N_4 and (14)-S_4, [91a,92a],
discussed above.

<u>45</u> <u>46</u>

Two Ru(IV) mono-oxo complexes have been characterised with the sterically
encumbered macrocycle, <u>46</u>, (tmc), [93b,93d]. The tertiary amine groups are
stronger σ-donors than secondary NH groups, so that the ligand is better able
to stabilise higher oxidation states of ruthenium. In the complex trans-
$[Ru(tmc)O(MeCN)]^{2+}$, the five and six membered rings have chair and gauche
conformations and the metal coordination geometry is distorted octahedral
[93b], with the Ru(IV) ion lying above the plane of the equatorial nitrogen
atoms and directed towards the axial nitrogen of the bound acetonitrile. The
N-Me groups are all arranged trans to the oxo group, with an (R,S,R,S) set of

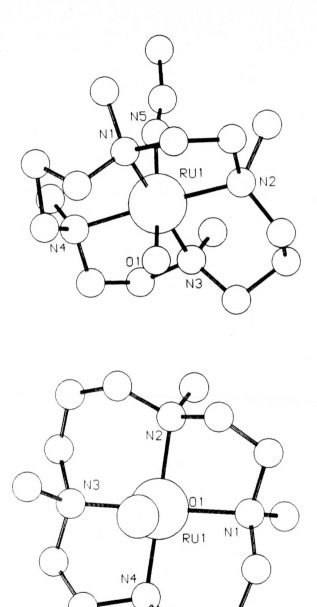

Figure 27. X-Ray crystal structures of the trans Ru(IV) complexes of <u>46</u> (tmc).
(a) [Ru(tmc)O(MeCN)]$^{2+}$ cation.
(b) [Ru(tmc)O(Cl)]$^{+}$ cation.

nitrogen configurations. Such an arrangement may be contrasted with the (R,S,R,R) array observed in the related monocation, trans-[Ru(IV)(tmc)O(Cl)]$^+$ in which three of the N-methyl groups are cis to the Ru=O group, (Figures 27a and 27b). Such an arrangement dictates that the two six-membered chelate rings have different conformations, one a twist-boat, the other a chair, although the 5-ring chelates remain gauche.

The ruthenium(VI) dioxo complex of 46 (tmc) i.e. trans-[Ru(tmc)O$_2$]$^{2+}$ is rather unstable both in solution and in the solid state [93f]. This instability may be associated with severe destabilising steric interactions between the axial N-Me groups and the oxygen lone-pairs. In order to alleviate this steric compression, equivalent complexes with the larger (15) and (16)-ring cycles, 47 and 48, were studied, [93e]. The crystal structure of the dicationic dioxoruthenium(VI) complex with 47 was rather disordered but

47

48

the Ru=O bond lengths were very similar to those found with 48, (1.72 compared to 1.71A^o). In the (16)-ring complex with 48, both independent (6)-ring chelates were in the chair conformation, and the N-methyl groups assumed the 'two-up, two-down' configuration, similar to that observed with the NH groups in the ruthenium(III) complex with cyclam, (Figure 26b).

Finally, the synthesis of osmium macrocyclic tertiary amine complexes has recently been described, [100]. Reaction of hexachloroosmate(IV) with either cyclam or tetramethyl-cyclam (TMC) under reducing conditions gave the trans-dichloroosmium(III) complexes which could be oxidised with peroxide to the trans-dioxoosmium(VI) dications (equation 1). Such a reaction sequence is closely analogous to that used to prepare the corresponding ruthenium complexes. In acidic solution, the trans-dioxo TMC complex exhibits a reversible three electron reduction wave, forming trans-[Os^{III}(TMC)(OH)$_2$]$^+$.

$$Na_2OsCl_6 + L \xrightarrow[\text{EtOH, } \Delta]{Sn} \text{trans-[OsLCl}_2]^+ \xrightarrow[\text{H}_2\text{O}]{H_2O_2} \text{trans-[OsLO}_2]^+ \quad \text{.....(1)}$$

11. REFERENCES

1 D.H. Busch, J.H. Cameron, N. Herron and G.L. Neer, J. Am. Chem. Soc.
   105 (1983) 298; D.H. Busch, G.G. Christoph, J.H. Cameron, W.R. Callahan,
   S.C. Jackels, D.J. Olszanski, J.J. Grzybowski, N. Herron and
   L.L. Zimmer, J. Am. Chem. Soc. 105 (1983) 6585.
2 M.J. Kappel, V.L. Pecoraro and K.N. Raymond, Inorg. Chem. 24 (1985)
   2447 and references therein.
3 A.M. Sargeson, Chemistry in Britain 15 (1979) 23; A.M. Sargeson,
   Pure Appl. Chem. 56 (1984) 1603; R.I. Geue, M.G. McCarthy,
   A.M. Sargeson, E. Horn and M.R. Snow, J. Chem. Soc. Chem. Commun. (1986)
   848.
4 'Copper Coordination Chemistry: Biochemical and Inorganic Perspectives'
   K.D. Karlin and J. Zubieta, Adenine Press, New York, 1983; D.E. Fenton
   in 'Advances in Inorganic and Bioinorganic Mechanisms' ed. A.G. Sykes,
   Academic Press, London, 1983, p. 187.
5 R. Louis, Y. Agnus and R. Weiss, Acta Cryst. B35 (1979) 2905.
6 J. Comarmond, B. Dietrich, J.M. Lehn and R. Louis, J. Chem. Soc. Chem.
   Commun. (1985) 74; G. Ferguson, C.R. Langrick, K.E. Matthes and
   D. Parker, J. Chem. Soc. Chem. Commun. (1985) 1609.
7 H.M. Colquhoun, J.F. Stoddart and D.J. Williams, Angew. Chem. Int. Ed.
   Engl. 25 (1986) 487.
8 J. Dale, Isr. J. Chem. 20 (1980) 3.
9 A.H. Alberts, J.M. Lehn and D. Parker, J. Chem. Soc. Dalton Trans. (1985)
   2311.
10 E.I. Solomon in 'Copper Proteins, Metal Ions in Biology', T.G. Spiro ed.,
   Wiley, New York, Vol. 3 (1981) pp. 41-108.
11 P.K. Coughlin and S.J. Lippard, J. Am. Chem. Soc. 103 (1981) 3328.
12 P.M. Collman, H.C. Freeman, J.M. Guss, M. Murata, V.A. Norris,
   J.A.M. Ramshaw and M.P. Venkatappa, Nature (London) 272 (1978) 319.
13 J. Comarmond, P. Plumeré, J.M. Lehn, Y. Agnus, R. Louis, R. Weiss,
   O. Kahn and I. Morgenstern-Badarau, J. Am. Chem. Soc. 104 (1982) 6330.
14 S.C. Rawle, J.R. Hartman, D.J. Watkin and S.R. Cooper, J. Chem. Soc. Chem.
   Commun. (1986) 1083.
15 E. Kimura, Coord. Chem. Rev. 15 (1986) 1.
16 R.O. Gould, A.J. Lavery and M. Schröder, J. Chem. Soc. Chem. Commun.
   (1985), 1492; E.R. Dockal, L.L. Diaddario, M.D. Glick and
   D.B. Rorabacher, J. Am. Chem. Soc. 99 (1977) 4530.
17 F.V. Lovecchio, E.S. Gore and D.H. Busch, J. Am. Chem. Soc. 96 (1974)
   3109.
18 A.M. Tait, M.Z. Hoffmann and E. Hayon, Inorg. Chem. 15 (1976) 934.
19 N. Jubran, G. Ginzburg, H. Cohen and D. Meyerstein, J. Chem. Soc. Chem.
   Commun. (1982) 517.
20 C.O. Dietrich-Buchecker, J.M. Kern and J.P. Sauvage, J. Chem. Soc. Chem.
   Commun. (1985) 760.
21 M. Ceserio, C.O. Dietrich-Buchecker, J. Guilhem, C. Pascard and
   J.P. Sauvage, J. Chem. Soc. Chem. Commun. (1985) 244.
22 J. Lewis and M. Schroder, J. Chem. Soc. Dalton Trans. (1982) 1085.
23 E. Kimura and M. Kodama, J. Chem. Soc. Dalton Trans. (1981) 694.
24 Y. Kushi, R. Machida and E. Kimura, J. Chem. Soc. Chem. Commun. (1985)
   216.
25 B. Bosnich, R. Mason, P.J. Pauling, G.B. Robertson and M.L. Tobe, Chem.
   Commun. (1965) 97; V.J. Thorn, C.C. Fox, J.C.A. Boeyens and
   R.D. Hancock, J. Am. Chem. Soc. 106 (1984) 5947.
26 T. Ito, M. Sugimoto, K. Toriumi and H. Ito, Chem. Lett. (1981) 1477.
27 R.D. Shannon, Acta Cryst. Sect. A 32 (1976) 751.
28 R. Louis, D. Pelissard and R. Weiss, Acta Cryst. B32 (1976) 1480.
29 R. Louis, Y. Agnus and R. Weiss, Acta Cryst. B33 (1977) 1418.
30 G. Ferguson, R. McCrindle and M. Parvez, Acta Cryst. C40 (1984) 354.
31 M.L. Campbell and N.K. Dalley, Acta Cryst. B37 (1981) 1750.
32 M.G.B. Drew, D.A. Rice and S.B. Silong, C40 (1984) 2014.

224

33 M.G.B. Drew, D.A. Rice and K.M. Richards, J. Chem. Soc. Dalton Trans. (1980) 2503.

34 D.E. Fenton, D.H. Cook, I.W. Nowell and P.E. Walker, J. Chem. Soc. Chem. Commun. (1978) 283.

35 H.W. Roesky, E. Reymann, J. Schimkowiak, M. Noltemeyer, W. Pinkert and G.M. Sheldrick, J. Chem. Soc. Chem. Commun. (1983) 981.

36 P.G. Jones, T. Gries, H. Grützmacher, H.W. Roesky, J. Schimkowiak and G.M. Sheldrick, Ang. Chem. Int. Ed. Engl. 23 (1984) 376.

37 T. Ito, H. Ito and K. Toriumi, Chem. Lett. (1981) 1101.

38 K.B. Mertes, Inorg. Chem. 17 (1978) 49.

39 J.H. Kim and G.W. Everett, Inorg. Chem. 20 (1981) 853.

40 C.H. Park, B. Lee and G.W. Everett, Inorg. Chem. 21 (1982) 1681.

41 R. Louis, D. Pelissard and R. Weiss, Acta Cryst. B30 (1974) 1889.

42 G. Ferguson, R. McCrindle, A.J. McAlees, M. Parvez and D.K. Stephenson, J. Chem. Soc. Dalton Trans. (1983) 1865; R. McCrindle, G. Ferguson, A.J. McAlees, M. Parvez and D.K. Stephenson, J. Chem. Soc. Dalton Trans. (1982) 1291.

43 M. Yamashita, H. Ito, K. Toriumi and T. Ito, Inorg. Chem. 22 (1983) 1566.

44 A.L. Balch, L.A. Fossett, M.M. Olmstead, D.E. Oram and P.E. Reedy, J. Amer. Chem. Soc. 107 (1985) 5272.

45 D. Parker, J. Rimmer and J.M. Lehn, J. Chem. Soc. Dalton Trans. (1985) 1517.

46 D. Parker, J. Chem. Soc. Chem. Commun. (1985) 1129, and unpublished work.

47 E. Kimura, Y. Lin, R. Machida and H. Zenda, J. Chem. Soc. Chem. Commun. (1986) 1020.

48 A.J. Blake, T.I. Hyde, R.S.E. Smith and M. Schroder, J. Chem. Soc. Chem. Commun. (1986) 334.

49 D.R. Alston, A.M.Z. Slawin, J.F. Stoddart and D.J. Williams, Angew. Chem. Int. Ed. Engl. 23 (1984) 821.

50 H.M. Colquhoun, S.M. Doughty, J.F. Stoddart and D.J. Williams, Angew. Chem. Int. Ed. Engl. 23 (1984) 235.

51 H.M. Colquhoun, J.F. Stoddart and D.J. Williams, J. Amer. Chem. Soc. 104 (1982) 1426.

52 H.M. Colquhoun, S.M. Doughty, A.M.Z. Slawin, J.F. Stoddart and D.J. Williams, Angew. Chem. Int. Ed. Engl. 24 (1985) 135.

53 A. Carroy, C.R. Langrick, J.M. Lehn, K.E. Matthes and D. Parker, Helv. Chim. Acta 69 (1986) 580.

54 B.A. Boyce, A. Carroy, J.M. Lehn and D. Parker, J. Chem. Soc. Chem. Commun. (1984) 1546.

55 J.P. Lecomte, J.M. Lehn, D. Parker, J. Guilhem and C. Pascard, J. Chem. Soc. Chem. Commun. (1983) 296.

56 P.K. Bhattacharga, J. Chem. Soc. Dalton Trans. (1980) 810.

57 N.W. Alcock, D.C. Liles, M. McPartlin and P.A. Tasker, J. Chem. Soc. Chem. Commun. (1974) 727.

58 M.G.B. Drewl, A.H.B. Othman, S.G. McFall and S.M. Nelson, J. Chem. Soc. Chem. Commun. (1977) 558.

59 M.M. Bishop, J. Lewis, T.D. O'Donoghue and P.R. Raithby, J. Chem. Soc. Chem. Commun. (1978) 476.

60 E. Fleischer and S. Hawkinson, J. Am. Chem. Soc. 89 (1967) 720.

61 S. Martin and D.H. Busch, Inorg. Chem. 8 (1969) 1859.

62 B.B. Hughes, R.C. Haltiwanger, C.G. Pierpont, M. Hampton and G.L. Blackmer, Inorg. Chem. 19 (1980) 1801.

63 V. McKee and W.B. Shepard, J. Chem. Soc. Chem. Commun. (1985) 158.

64 J. Ferguson and M.L. Tobe, Inorg. Chim. Acta 4 (1970) 109.

65 E. Bang and O. Monsted, Acta Chem. Scand. A36 (1982) 353.

66 E. Bang and O. Monsted, Acta Chem. Scand. A38 (1984) 281.

67 R. Temple, D.A. House and W.T. Robinson, Acta Cryst. C40 (1984) 1789.

68 B.A. Nair, T. Ramasami and D. Ramaswamy, Inorg. Chem. 25 (1986) 51.

69 D. Yang and D.A. House, Inorg. Chim. Acta 64 (1982) L167.

70 D.A. House, R.W. Hay and N.A. Ali, Inorg. Chim. Acta 72 (1983) 239.

71   R.G. Swisher, G.A. Brown, R.C. Smierciak and E.L. Blinn, Inorg. Chem.
     20 (1981) 3947.
72   R.E. DeSimone and T.M. Tighe, J. Inorg. Nucl. Chem. 38 (1976) 1623.
73   R.E. DeSimone and M.D. Glick, J. Am. Chem. Soc. 97 (1975) 942.
74   R.E. DeSimone and M.D. Glick, J. Coord. Chem. 5 (1976) 181.
75   R.E. DeSimone and M.D. Glick, J. Am. Chem. Soc. 98 (1976) 762.
76   C.G. Kuehn and S.S. Isied, Prog. Inorg. Chem. 27 (1980) 153.
77   R.E. DeSimone and M.D. Glick, Inorg. Chem. 17 (1978) 3574.
78   R.E. DeSimone, J. Cragel, W.H. Usley and M.D. Glick, J. Coord. Chem. 9
     (1979) 167.
79   D. Sevdic and L. Fekete, Polyhedron 4 (1985) 1371.
80   M.W. Anker, J. Chatt, G.J. Leigh and A.J. Wedd, J. Chem. Soc. Dalton
     Trans. (1975) 2639.
81   A.D. Weettand and N. Mariithi, Inorg. Chem. 11 (1972) 2971.
82   D. Sevdic, L. Fekete and H. Meider, J. Inorg. Nucl. Chem. 42 (1980) 885.
83   D. Sevdic and H. Meider, J. Inorg. Nucl. Chem. 43 (1981) 153.
84   J. Simon, A.R. Ketring, D.E. Troutner, W.A. Volkert and R.A. Holmes,
     Radiochem. Radioanal. Lett. 38 (1979) 133.
85   D.E. Troutner, J. Simon, A.R. Ketring, W.A. Volkert and R.A. Holmes,
     J. Nucl. Med. 21 (1980) 443.
86   J. Simon, D.E. Troutner, W.A. Volkert and R.A. Holmes, Radiochem.
     Radioanal. Lett. 47 (1981) 111.
87   S.A. Zuckman, G.M. Freeman, D.E. Troutner, W.A. Volkert, R.A. Holmes,
     D.G. Van der Weer and E.K. Barefield, Inorg. Chem. 20 (1981) 2386.
88   P.A. Tasker and L.J. Sklar, Cryst. Mol. Struct. 5 (1975) 329.
89   C.K. Fair, D.E. Troutner, E.O. Schlemper, R.K. Murmamn and M.L. Hoppe,
     Acta Cryst. Sect. C C40 (1984) 1544.
90   S. Jurisson, E.O. Schlemper, D.E. Troutner, L.R. Canning, D.P. Nowotrik
     and R.D. Neirinckx, Inorg. Chem. 25 (1986) 543.
91a) T.F. Lai and C.K. Poon, J. Chem. Soc. Dalton Trans. (1982) 1465.
  b) C.K. Poon and C.M. Ché, J. Chem. Soc. Chem. Commun. (1979) 861.
  c) A.J. Blake, T.I. Hyde, R.S.E. Smith and M. Schroder, J. Chem. Soc. Chem.
     Commun. (1986) 334.
92a) C.M. Ché, S.S. Kwang, C.K. Poon, T.F. Lai and C.W. Mak, Inorg. Chem.
     24 (1985) 1359.
  b) C.M. Ché, S.S. Kwong and C.K. Poon, Inorg. Chem. 24 (1985) 1601.
  c) D. Walker and H. Taube, Inorg. Chem. 20 (1981) 2828.
93a) C.M. Ché, K.Y. Wong and C.K. Poon, Inorg. Chem. 25 (1986) 1809.
  b) C.M. Ché, K.Y. Wong and C.W. Mak, J. Chem. Soc. Chem. Commun. (1985) 546.
  c) C.M. Ché, T.W. Tang and C.K. Poon, J. Chem. Soc. Chem. Commun. (1984) 641.
  d) C.M. Ché, K.Y. Wong and C.W. Mak, J. Chem. Soc. Chem. Commun. (1985) 988.
  e) C.W. Mak, C.M. Ché and K.Y. Wong, J. Chem. Soc. Chem. Commun. (1985) 986.
  f) C.M. Ché, K.Y. Wong and C.K. Poon, Inorg. Chem. 24 (1985) 1797.
  g) C.M. Ché and K.Y. Wong, J. Chem. Soc. Chem. Commun. (1986) 229.
94   C.K. Poon and C.M. Ché, J. Chem. Soc. Dalton Trans. (1981) 495.
95   K. Travis and D.H. Busch, Inorg. Chem. 13 (1974) 2591.
96   T.F. Lai and C.K. Poon, Inorg. Chem. 15 (1976) 1562.
97   P.K. Chan, D.A. Isabirye and C.K. Poon, Inorg. Chem. 14 (1975) 2579.
98   S.S. Issied, Inorg. Chem. 19 (1980) 911.
99   C.K. Poon and C.M. Ché, J. Chem. Soc. Dalton Trans. (1981), 1336.
100 C.M. Ché and W.K. Cheng, J. Am. Chem. Soc. 108 (1986) 4644.

226

The figures of this article are reproduced, or redrawn, with permission, from:

Fig. 1. J. Chem. Soc., Chem. Commun. (1985) 216.
     2. Acta Crystallogr., B32 (1976) 1480.
     3. Acta Crystallogr., B33 (1977) 1418.
     4. Acta Crystallogr., C40 (1984) 354.
     5. Acta Crystallogr., C40 (1984) 2014.
     6. Angew. Chem., Int. Ed. Engl., 23 (1984) 376.
     7. Chem. Lett. (1981) 1101.
     8. Inorg. Chem., 20(3) (1981) 853.
     9. Acta Crystallogr., B30 (1974) 1889.
    10. J. Chem. Soc., Dalton Trans. (1983) 1865.
    11. J. Chem. Soc., Dalton Trans. (1983) 1865.
    12. J. Am. Chem. Soc. (1985) 5272.
         Copyright American Chemical Society
    13. J. Chem. Soc., Chem. Commun. (1983) 296.
    14. J. Chem. Soc., Chem. Commun. (1974) 727.
    15. J. Chem. Soc., Chem. Commun. (1977) 558.
    16. J. Chem. Soc., Chem. Commun. (1978) 476.
    17. Inorg. Chem., 19(6) (1980) 1801.
    18. J. Chem. Soc., Chem. Commun. (1985) 158.
    19. Acta Chem. Scand., A36 (1982) 353.
    20. Acta Chem. Scand., A38 (1984) 281.
    21. Inorg. Chem., 17(10) (1978) 2885.
    22. J. Coord. Chem., 9 (1979) 167.
         Copyright Gordon and Breach Science Publishers Inc.
    23. Inorg. Chem., 17(10) (1978) 3574.
    24. Inorg. Chem., 20(8) (1981) 2386.
    25. J. Chem. Soc., Dalton Trans. (1982) 1465.
    26. (a) Inorg. Chem., 24(9) (1985) 1359.
         (b) Inorg. Chem., 20(9) (1981) 2828.
    27   (a) J. Chem. Soc., Chem. Commun. (1985) 546.
         (b) J. Chem. Soc., Chem. Commun. (1985) 988.

SUBJECT INDEX

(grouped by chapter)

Index for Chapter 1

Index for Chapter 2

Index For Chapter 3

244

X-Ray Structures Of (Continued)